Design Thinking in the Middle Grades

Design Thinking in the Middle Grades

Transforming Mathematics and Science Learning

Reagan Curtis, Darran R. Cairns,
and Johnna J. Bolyard

Foreword by James W. Stigler

TEACHERS COLLEGE PRESS

TEACHERS COLLEGE | COLUMBIA UNIVERSITY
NEW YORK AND LONDON

Published by Teachers College Press,® 1234 Amsterdam Avenue, New York, NY 10027

Copyright © 2023 by Teachers College, Columbia University

Parts of this work were supported by the National Science Foundation through the "Research Experience for Teachers: Energy and the Environment" project (NSF#0908582) and by the West Virginia Department of Education Through the Math Sciences Partnership, "Project TESAL: Teachers Engaged in STEM and Literacy."

Front cover by adam bohannon design. Image by Robert Herschede via Flickr creative commons.

Library of Congress Cataloging-in-Publication Data is available at loc.gov

ISBN 978-0-8077-6780-1 (paper)
ISBN 978-0-8077-6781-8 (hardcover)
ISBN 978-0-8077-8145-6 (ebook)

Printed on acid-free paper
Manufactured in the United States of America

This work is dedicated to the many teacher colleagues who have shared their expertise and challenges with us over the years.

Contents

PART III: MAKING IT YOUR OWN

Foreword

Many recent "advances" in applications of technology to education require a more and more atomized view of what learning is. If the goals of learning can be stated as micro-objectives, and if students' learning can be assessed separately for each of these bits of knowledge and skill, then modern algorithms of machine learning and AI can be used to order and present to students just the opportunities they need to learn all the stuff necessary to participate in the modern knowledge economy.

But there's a fly in this ointment. It may appear tiny at first, but on further reflection it's huge. As teaching and learning become more atomized, we increasingly see students who have mastered the bits yet have little understanding of the overall domain. They do fine on the test, provided the bits are assessed "fairly" and in accordance with how they were taught. But their knowledge fails to transfer to novel situations.

Transfer of knowledge requires connections. The bits need to be connected with one another and to core concepts that cut across the domain to which we hope knowledge will transfer. Concepts need to be connected to important representational systems, such as the equations and algebraic notation scientists use to help them understand complex systems. And concepts, skills, and representations need to be connected to the world and to people's lives in a flexible way.

Transfer requires more than just a mastery of the bits. It requires the learner to be able to take all the bits of knowledge they have acquired and then to coordinate them and adjust them to achieve an important goal or solve an important problem. Connections are important because they guide and open up new possibilities for the problem-solver, helping them find and apply bits of knowledge originally learned in one domain to another, perhaps unrelated, domain. As these bits come to the fore, they often must be modified or combined, creatively, to fit a new situation.

Many learning scientists still focus their efforts on helping students learn the bits more quickly and efficiently. They live in what some have referred to as the "Zone of Wishful Thinking," hoping, in the face of all evidence to the contrary, that students will be able to retain and apply their knowledge of the bits, later, in new situations.

Refreshingly, the authors of this book have taken an almost opposite path. Instead of teaching the bits and hoping for the best, they are starting out with challenges in the world and then working backward, helping students develop the knowledge, skills, connections, dispositions, and creativity they will need to meet these challenges. They are putting the focus on the end goal—transfer—and then teaching the bits in ways that will contribute to students' progress toward this ambitious goal.

The authors rely in their work on the increasingly popular disciplines of design thinking and systems improvement. Although they aren't the first to apply design thinking methodologies to the development and improvement of STEM curricula, their approach is unique and highly innovative. The design process is not just a means of developing educational resources, but is also an explicit and integral part of the program they are developing.

The design thinking process is not something researchers develop, teachers implement, and students benefit from. It is, in this book, something researchers, teachers, and students work on together. Students, in the end, not only learn the things they need to know based on current STEM content standards, but also walk away as practitioners of a disciplined approach to activating and using what they know to solve real problems in the world.

I hope these ideas, and this approach, pick up steam here in the United States. Although the work is at an early stage, it gives me hope. When we tire of trying to automate learning, and see the limitations more clearly of just mastering the bits of knowledge, I expect this is the kind of work that will rise in prominence. I'm looking forward to that day.

James W. Stigler
August 10, 2022
Los Angeles, CA

Acknowledgments

We could not have completed this work without our supportive family, friends, and professional networks, including the teacher colleagues with whom we developed and refined our approach. Shar provided endless motivation and fuel (good food) for our writing, as well as much-needed fun and distraction with her partner in crime, Tori, who first suggested a roller coaster for the cover of this book. We appreciate James W. Stigler for his thoughtful Foreword, as well as Patrick Hobb, Colleen Mount, Alison Silva, and peer reviewers who provided feedback on early chapters. We thank everyone at Teachers College Press who helped make this book a reality, especially Emily Spangler and Mike Olivo for insightful editing and helping us navigate the publication process.

Design Thinking in the Middle Grades

Introduction

We wrote this book for all middle school teachers who focus on science, technology, engineering, and mathematics, as well as for anyone who helps teachers improve teaching and learning in preservice teacher preparation or with inservice teachers. The middle grades are a significant time in students' educational development. This is a time when many students make critical decisions about what it means to be someone who "does" science, technology, engineering, and mathematics, and whether they fit into that role. Such decisions have serious impacts on their future education and career choices. Unfortunately, too many students determine in the middle grades that science, technology, engineering, and mathematics are not for them. We believe that this is, in part, a result of how the teaching and learning of these disciplines are experienced in the middle grades. These disciplines are often taught separately, with few opportunities to understand how they can work together to solve real, meaningful problems that are relevant to students' lives. As an alternative, we share an approach that highlights the integrated nature of these disciplines and their relevance to students' lives.

In very brief terms, the core of our approach is engaging learners to pursue solutions to challenges in their everyday world through design-based learning cycles utilizing mathematical modeling to analyze alternate solution pathways, predict which potential solutions may be successful, and test actual performance against predictions. Before continuing, clarifying the relations among "design-based" and the better known "project-based" and "problem-based" teaching and learning may be helpful (see Figure I.1). Design-based learning is a special case, or subset, of a broader approach known as problem-based learning and often is also project based. Project-based learning contextualizes content in thematic units, or larger projects such as exploring an ecological system or energy efficiency. Problem-based learning is focused on student-generated problems and learning that occurs as teachers guide students seeking solutions to those problems. Problem-based learning is often contextualized in thematic units, as in project-based learning. That makes design-based learning a form of both project-based and problem-based learning that explicitly uses the design process and

Figure I.1. Relations Among Project-Based, Problem-Based, and Design-Based Learning

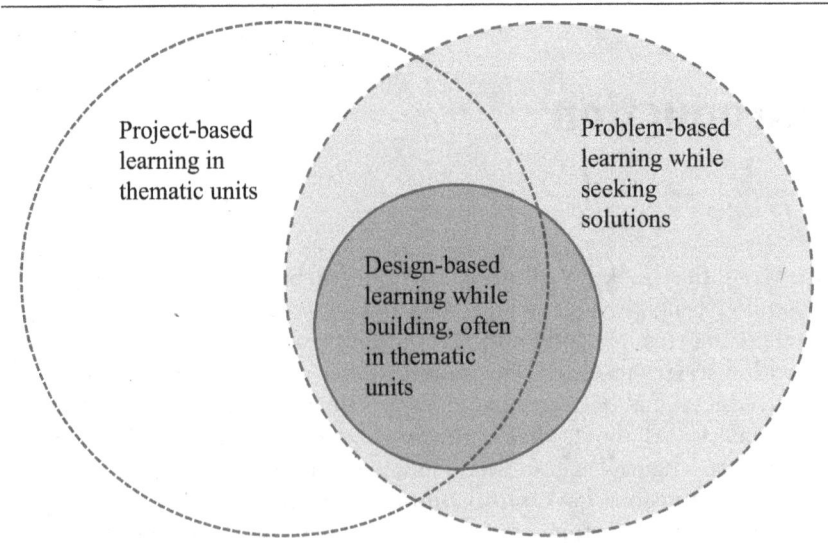

involves engineering artifacts, processes, and mathematical models as (at least part of) the solution. In our design-based approach, we emphasize connecting grade-appropriate science and mathematics content standards and practices to these everyday challenges, and integrating literacy content standards and practices with evidence-based argument through design briefs, presentations, and reflection.

You are almost certainly familiar with the acronyms STEM and STEAM, and may have wondered why we do not use them to describe our approach. *STEM* (science, technology, engineering, and mathematics) and later *STEAM* (integrating the arts) were introduced to publicize two core ideas. The first is that each of the disciplines that make up the acronym should have a higher profile in education. We agree, and efforts to raise their profile have been largely successful. The second core idea is that meaningful integration across these areas is critical to support deep learning and should have a more dominant focus in education. Again, we agree, but the penetration of this idea into dominant educational practice and research has been less successful. Unfortunately, we find far too many examples of work labeled as STEM or STEAM that does not make it clear which of the disciplines are being integrated and how or even if they are integrated in meaningful ways at all. For that reason, we prefer to identify our work with the disciplines themselves, describing what we do as integrating mathematics, engineering, science, and literacy.

The particular configuration of theory and evidence-based practices we advocate for in this book did not spring fully formed out of thin air. We developed our approach and gathered evidence of its effectiveness collaboratively with current teachers participating in professional development programs with us, drawing on what we know from research on teaching and learning, what our teacher colleagues told us about their experiences as learners and with their students, and continuously (re)designing, testing, and refining over the last decade. As we worked with teachers, we were struck by how effective teachers found this approach to be for themselves and their students, as well as by how few high-quality resources were available to support their learning more deeply about and applying the concepts we describe in this book.

Who are we? We are a heterogeneous group of faculty focused on developing and sharing approaches to improving mathematics and science teaching and learning. We collectively strive to embody and model deep content-area expertise in science, mathematics, engineering, and education; strong theoretical grounding in empirical research on teaching and learning pedagogy and evidence-based practices; and an enduring commitment to fostering equity in schools and beyond.

Johnna Bolyard (Ph.D. in Mathematics Education from George Mason University) is an associate professor in the School of Education at West Virginia University, where she teaches undergraduate mathematics methods and content courses for teachers and graduate-level courses in mathematics education research and mathematics leadership. Johnna brings deep expertise in mathematics content and mathematics education pedagogy to support developing and practicing teachers. Her research focus is on the development of pre- and inservice mathematics teaching practice. In particular, she explores how teachers take up teaching practices that support the mathematical development of all students.

Darran Cairns (Ph.D. in Materials Science and Engineering from University of Birmingham, UK) is faculty in mechanical engineering at the University of Missouri—Kansas City, where he teaches graduate and undergraduate engineering and product design courses and is director of the School of Science and Engineering High School Math Academy. Darran brings deep engineering, science, and pedagogical expertise to supporting learners, as well as continuing a productive research agenda in materials science and engineering.

Reagan Curtis (Ph.D. in Educational Psychology from the University of California, Santa Barbara) is Chester E. & Helen B. Derrick Endowed Professor in the School of Education and Director of the Program Evaluation and Research Center at West Virginia University. Reagan works within a learning sciences perspective to focus on the evidence base for teaching and learning practices and within an evaluation research framework to develop and improve professional development programs.

Our academic credentials are less important than the experience we gained working with teacher colleagues in and out of classrooms to support teacher and student learning. We began this journey, although the seeds were there long before, in 2009, when Darran and Reagan received funding from the National Science Foundation to bring teachers into engineering labs as research partners and to use those experiences to enhance their classroom teaching practice. This project was grounded in energy and the environment as a central theme, and we worked with teachers to make meaningful connections and to implement curricular activities in their classroom instruction, building on their experiences at the university in engineering research labs.

This was prior to engineering and design having the prominence they do now in national science standards documents, and we sometimes struggled to make explicit connections between teachers' research experience in university labs and their teaching practice in their own classrooms. This challenge was most evident for mathematics teachers, and yet one of the more successful examples of such a connection came from two mathematics teachers who collaborated first on a module around fuel efficiency related to tire inflation and then on a module around insulation and heating/cooling efficiency. As mathematics teachers, they were less comfortable with hands-on experiments and building of designs, but were quite comfortable with the mathematics content needed to predict and test the efficiency of student designs (e.g., geometry of three-dimensional shapes such as tires, calculating slope, taking real-world measurements, and characterizing variability). Working together, they taught integrative lessons and found that their students more deeply understood the content standards and objectives tied to those lessons than had past classes. In fact, these teachers found they were able to teach required content standards and objectives more quickly in integrative units as compared to prior years, leaving additional time later in the year to review and prepare for standardized testing.

This experience with mathematics-forward–design-based learning was transformative for these teachers, and their excitement to continue refining and extending their use of design-based learning was infectious. We have seen similar successful implementation of science-forward–design-based learning and will provide examples later in this book. Single-subject (mathematics or science) teachers experiencing success with this approach often, although not always, reach out and grow their professional networks to include cross-disciplinary colleagues, leading to increasingly integrative approaches to curricula in their schools. They also internalize design thinking and apply it to their own teaching practices and daily lives. This helped us recognize the applicability of the design process to the development, testing, and revision of instructional practices by teachers (and indeed ourselves).

In 2012, we were fortunate to have Johnna join our group as we launched a math-science partnership that would bring to teachers a model of professional development that focused on mathematics-science-literacy content standards integration through design-based learning. It was through this project that we developed the integration of the components we describe in Chapters 2 and 3, the evidence base for our approach described in Chapter 4, and many of the curricular resources we describe in the remainder of this book. We continue to work with teacher colleagues with whom we developed relationships through these projects and seek opportunities to refine and share what we are learning more broadly.

We have organized this book into three sections. The five chapters in Part I set the stage to help you understand our approach. Chapter 1 introduces the core components of our approach, and Chapter 2 delves deeper into those core components. Chapter 3 discusses two teaching challenges our approach helps address (i.e., fostering equity and supporting productive struggle) and important considerations that help the approach be most successful (i.e., complex instruction and reflective practice). Chapter 4 presents evidence for the effectiveness of this approach. Chapter 5 explores how integrating these components aligns with the best practices described in national standards and research on teaching and learning, as well as how to think about individualizing your practice with your students while implementing our approach.

Part II focuses on making the approach more visible with concrete descriptions of lesson plans and teachers' experiences implementing and refining those lessons. Chapters 6 through 10 focus on various practical considerations to support your ability to individualize your use of this approach: designing processes versus designing artifacts, designing with many versus few constraints, designing with high versus low amounts of learner supports or scaffolding, designing in different grade levels and content areas, and designing in different instructional modalities (i.e., in-class, at home, online).

Part III brings it all together to help you make what you learn here your own, refining it for your learners and teaching context. Chapter 11 focuses on how to link an integrative series of design-based lessons into larger thematic curricular chunks. Chapter 12 focuses on how to create your own design-based lessons for specific content standards and objectives, including how to "fix" material from traditional textbooks to become a foundation for design-based learning activities.

Part I

UNDERSTANDING THE APPROACH

CHAPTER 1

Thinking Differently About Teaching and Learning

You have picked up this book because you want to expand how you think about and support mathematics and science learning in the middle grades. You may be familiar with terms such as "problem-based" or "project-based" learning, but how they connect to or are different from engineering design is likely less familiar. A key focus of this book is understanding how to utilize engineering design cycles for learning, an approach we refer to as "design-based learning." While there are many aspects that set design-based learning apart, the central idea is that students use the design process to build and test solutions to real-world problems. We think a great way to begin to understand this distinction and the approach we advocate in this book is to hear about how one of our teacher colleagues discovered the powerful effect design-based learning could have on his students. While names have been changed, what follows is a real-life example.

WHAT IT LOOKS LIKE IN A REAL CLASSROOM

Mr. Shaker walked into his 6th-grade class excited to try out what he had been learning about design-based lessons. While many of his students were progressing well, there were a few he could not yet fully engage in the learning activities he used each year. They would hang back during group work and let others take the lead. As a special education teacher focused on mathematics and science, he knew some of his students' records had led other teachers not to push them or expect too much. Further, he had noticed that *all* of his students had difficulty when asked to go beyond following a step-by-step process to a predetermined solution. Many of his students did not seem to have confidence that their own thinking could be important or that their experiences outside of school could be relevant to learning in school. He was ready to try something new.

Mr. Shaker told his students they would begin focusing on how to improve everyday things in people's lives. Along the way they would pick up the math and science they needed to put their solutions into action. While these would be the same math and science concepts schools wanted them to know, the most important thing was that the math and science they learned would actually help them create solutions to problems they cared about. They would learn to share their solutions and evidence of their effectiveness with others in writing and through presentations.

He asked how many students liked to eat hard-boiled eggs. He also asked if some eggs tasted better than others. He then introduced a real-world problem: discolored yolks are a problem in the food service industry as well as for making students' own breakfasts. He asked if they thought a good place to start could be figuring out how to make "the perfect" hard-boiled egg with a perfect solid yellow yolk and no discoloration. He then asked students to go home and, if possible, find someone who would help them think through how to do that. After returning to Mr. Shaker's classroom, students worked in groups to boil multiple eggs and develop their own best process; created trifold brochures and a manual describing their process for a perfect hard-boiled egg; produced a second perfect egg following another group's manual; and then wrote an essay comparing, contrasting, and reflecting upon their process and instruction manual as well as that of another group. The activity lasted a week.

Students considered the transfer of heat from the stove to the water and the egg; explored physical and chemical changes in the egg during the boiling process; measured and summarized numerical data with units of time for multiple stages of their process and distance for the thickness of discolored regions in the eggs; integrated their quantitative results and instructions into a text; compared and contrasted information from experiments; and used the design process. Both science and mathematical principles were authentically explored through the design process, and the instruction manual allowed peer review and reflection on students' communication. For Mr. Shaker it was the level of student engagement that stood out to him.

> Students went home and asked their parents and grandparents to teach them to boil an egg and practiced with them. They then sat in the classroom carefully watching the water boil, timing every step. I had never seen that level of engagement with this group of students. It has made me a believer.

Emboldened by his students' response, Mr. Shaker next drew a challenge from his own life. His students knew he loved taking part in triathlons. He asked if they ever got in trouble for tracking mud into the house, and he told them how his wet, muddy shoes in the trunk of the car bothered his partner. Could they think of ways to help? He informed his students that he was not just looking for ONE solution; he wanted lots of possible solutions. Everyone's ideas were needed. Everyone's minds mattered.

Students agreed that if Mr. Shaker could not hose off or clean up his shoes before getting to the car, one solution might be a container for his shoes. But he did not have much money to spend and there was not much space in his trunk. He happened to have a bunch of cardboard boxes that had been broken down so that each was a single flat piece. His students worked in teams to plan possible designs. Students determined what volume was needed by measuring the size of pairs of sneakers, calculating and estimating for themselves the dimensions needed for the finished shoe boxes. They discussed how to figure out which designs met the "design constraints and criteria," meaning they could be made from a single sheet of cardboard, fit his shoes, and take up as little space in his trunk as possible.

Students sketched and built initial prototypes from *nets* (i.e., two-dimensional representations of three-dimensional objects) drawn to scale using pieces of paper; they cooperated and collaborated to ask questions; they redesigned their prototypes based on how they turned out and their earlier volume calculations; and then they built final scale models from their cardboard and presented their designs. Two students who rarely engaged in traditional learning tasks designed and built an innovative triangular prism. After this experience, Mr. Shaker reflected, "Whenever they are struggling in class now, I remind them of their success on the shoebox project. I am working on design projects to use in the spring semester with them now."

Mr. Shaker focused on mathematics, science, and literacy content standards and integrated them seamlessly to engage all of his students in meaningful integrative learning (see Figure 1.1). The standards we refer to are the Next Generation Science Standards (NGSS; NGSS Lead States, 2013; https://nap.nationalacademies.org/catalog/18290/next-generation-science-standards-for-states-by-states), Common Core Math Content Standards (CCSS-M; National Governors Association & Council of Chief State School Officers, 2010; http://www.corestandards.org/Math/), and Common Core English Language Arts Standards (CCSS-ELA; National Governors Association & Council of Chief State School Officers, 2010; http://www.corestandards.org/ELA-Literacy/).

Mr. Shaker's lesson was designed with both content and practice in mind and combined mathematical modeling and design-based learning to provide opportunities to explore mathematics content standards related to the volume of three-dimensional shapes and a science content standard related to collaborating during scientific investigations. In this example, Mr. Shaker's students produced a *design report*, which details the constraints, the steps in the design process, a prototype, the construction process, evaluation of the effectiveness of the product, and suggestions for redesign options. Mr. Shaker used the design report to support students in developing both literacy skills and enabling reflection on both the products they developed and the design process they used. This is one of many stories from our teacher colleagues as they have gained confidence in using a design-based learning approach.

Figure 1.1. Standards in the Shoebox Activity

CCSS-M Mathematics Content and Practice Standards

- 6.GA.2: Find the volume of a right rectangular prism with fractional edge lengths by packing it with unit cubes of the appropriate unit fraction edge lengths and show that the volume is the same as would be found by multiplying the edge lengths of the prism. Apply the formulas $V = lwh$ and $V = Bh$ to find the volumes of right rectangular prisms with fractional edge lengths in the context of solving real-world and mathematical problems.
- 6.GA.4: Represent three-dimensional figures using nets made up of rectangles and triangles, and use the nets to find the surface area of these figures. Apply these techniques in the context of solving real-world and mathematical problems.
- MP1: Make sense of problems and persevere in solving them.
- MP4: Model with mathematics.

NGSS Science Standards

- MS-ETS1-2: Define the criteria and constraints of a design problem with sufficient precision to ensure a successful solution, taking into account relevant scientific principles and potential impacts on people and the natural environment that may limit possible solutions.
- MS-ETS1-2: Evaluate competing design solutions using a systematic process to determine how well they meet the criteria and constraints of the problem.

CCSS-ELA Standards

- W.6.2: Write informative/explanatory texts to examine a topic and convey ideas, concepts, and information through the selection, organization, and analysis of relevant content.

WHY THIS BOOK?

The approach we share in this book is not *only* for teachers who focus on integrative science, technology, engineering, and mathematics instruction. This approach can help any mathematics or science teacher, as well as anyone who helps teachers improve teaching and learning in preservice teacher preparation or engages in professional development with current teachers. While we focus on the middle grades, our approach can be scaled for use in elementary and high school as well. In this first chapter, we hope to spark your interest and give you an overview of the approach we advocate for in this book. Our overall goal with this book is to support your successful use of our approach with your students.

Remember how Mr. Shaker integrated mathematics and science content and connected that content to real-world problems with design thinking. Students were able to use experiences from outside the classroom and connect those experiences to mathematics and science content while developing solutions to concrete challenges relevant to their lives. The middle grades are a critical time to engage the next generation so that they are prepared to solve tomorrow's challenges. Middle grade students often develop deep-seated understandings about what it means to do science and mathematics and whether their voices are valued in science and mathematics conversations. The understandings they develop in middle school strongly influence which education and career paths they believe are open to them. Too often, mathematics and science are taught in isolation from each other and are not connected to meaningful problems that matter to students. This isolated approach, combined with narrow beliefs about who can or should do science and mathematics, prevents many students from engaging fully in these disciplines. We have come to see the integration of theory and evidence-based practices that constitute our approach as a transformative new way of teaching and learning that is particularly important for the middle grades.

The core of our approach is engaging learners to pursue solutions to challenges in their everyday world through design-based learning cycles utilizing mathematical modeling to analyze alternate solution pathways, to predict which potential solutions may be successful, and to test actual performance against predictions (see Figure 1.2). We describe *mathematical modeling* more fully in later chapters, but a good working definition is "using mathematics or statistics to describe (i.e., model) a real-world situation and deduce additional information about the situation by mathematical or statistical computation and analysis" (Common Core State Standards Writing Team, 2013, p. 15). We emphasize connecting grade-appropriate science and mathematics content standards and practices to everyday challenges, and integrating literacy content standards and practices with evidence-based argument through design reports, presentations, and reflection.

Figure 1.2. Design, Mathematical Modeling, and Content Standards as Core Components

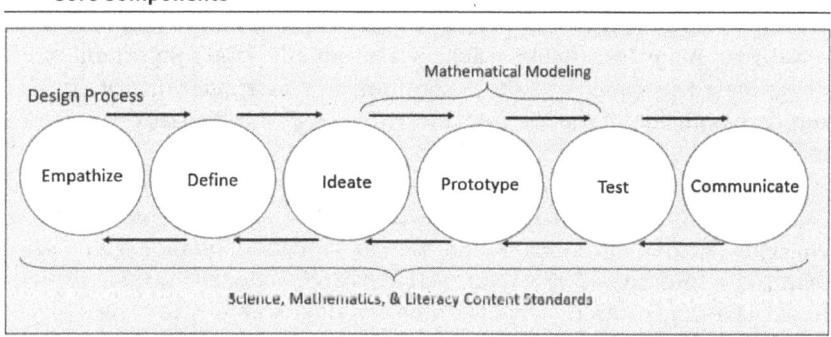

As you look at Figure 1.2, we hope you think about how our example from Mr. Shaker's classroom connects to this model. Recall how Mr. Shaker engaged his students in challenges from everyday life (i.e., boiling eggs, dirty shoes). Science content was integrated as students tried to understand why eggs solidify when boiled while other liquids do not and what happens when they boil too long. Mr. Shaker's students used mathematical modeling when they measured shoes and used those measurements to predict the volume their container designs would need to hold. They compared that volume to volume calculations from different design plans to predict which designs would work as students decided what design to actually build. Mr. Shaker integrated mathematical content as students learned about nets, connecting measurements of two-dimensional cardboard sheets to three-dimensional shoebox shapes. Literacy was integrated as students wrote about and presented the solutions they designed to the class and reflected on what they learned in the process.

National education standards documents make it clear that mathematics is an essential tool for scientific inquiry, and science is a critical context for developing mathematics competence (Doerr & English, 2006; National Research Council, 2006). Mutually reinforcing science and mathematics understandings while teaching either discipline is an important and effective interdisciplinary opportunity (Center for Educational Policy, 2007; National Council of Teachers of Mathematics [NCTM], 2000). Integrating literacy standards and practices further extends and grounds this opportunity for preparing students to meet tomorrow's challenges.

A Framework for Science Education gives engineering and technology a greater focus than prior science standards documents (Czerniak, 2007; NGSS Lead States, 2013). While middle grade content standards now include engineering and design in the science framework, they did not when we began this work, and the design process is not easy to learn. This is at least partially because design is a dynamic and iterative process rather than a specific skill or piece of content knowledge. Such dynamic processes have less often been part of traditional teacher certification programs, and the emphasis on engineering and design standards in middle grades occurred after most current teachers' preservice preparation. In this light, it makes sense that many practicing teachers are initially less comfortable with design-based learning and the integration in this framework of content standards outside of their certification area (e.g., mathematics for science teachers and vice versa).

In the approach we expand on in Chapter 2, mathematics, science, and engineering content standards and practices are integrated with cross-cutting concepts focused on systems and system models (Common Core State Standards Initiative, 2011; National Research Council, 2012). Design-based learning projects provide extensive opportunities for engaging in practices common to mathematics and science standards and framework documents while teaching and learning mathematics and science: defining problems, constructing explanations, developing models, and attending to precision.

As we implement this approach with teachers and in classrooms, we draw on evidence-based practices in an expanded model, described in Chapter 3, for structuring group learning that fosters equity in the classroom by ensuring that the voices of all group members are honored and that all learners contribute to meaningful tasks (Boaler, 2006). This is more challenging than it may seem, as practicing teachers will affirm. Yet it has the power to reframe who is contributing in classrooms and in what ways, thus, broadening views of who can do science and mathematics.

One key to fostering equity in the classroom is to focus on which students have *status* as contributors in the classroom and actively work to give status to students whose voices are heard less often, or not at all. Another key is to utilize tasks that are *group-worthy*: open-ended, complex tasks that provide multiple entry points, require interdependence among group members, and have clear criteria to evaluate group functioning (Lotan, 2003). We discuss this more deeply in later chapters, and there are excellent resources available through Stanford's Graduate School of Education (https://complexinstruction.stanford.edu/).

A critical understanding that we hope will come through more deeply as you read the rest of this book is that design thinking is supremely adaptable. It is a process that applies authentically to design-based learning projects in and out of classrooms as well as to the design and redesign of teachers' instructional practices and professional development. It applies just as much to designing solutions to everyday problems as it does to the National Academy of Engineering's (NAE, 2008) "grand challenges" (e.g., provide energy from fusion, engineer better medicines) and to "wicked problems" that the world faces today (e.g., increasing social justice through entrepreneurship, equitable and effective education for all; Buchanan, 1992; Rittel & Webber, 1973; http://www.engineeringchallenges.org/challenges.aspx; www.wickedproblems.com). We work to model and scaffold this mindset for and with our teacher colleagues. We have been gifted opportunities to see them do the same with us and with their students, and we hope that learners so enabled will contribute to the next generation of solutions the world so desperately needs.

CHAPTER 2

Engineering Design, Mathematics, Science, and Literacy

We hope that you are intrigued by the overview in Chapter 1 and motivated to learn more about how design, mathematical modeling, and math-science-literacy content integration can transform your teaching and learning practice. The design-based learning integrative framework we describe in this book is likely a new approach to instruction for you and your students. Initially, it will feel challenging. We have designed and organized this book to support your success. In this chapter, we cover the core components of our model. The next chapter will build on these core components to provide a full model of the approach we advocate. Later chapters focus on the evidence base for the effectiveness of this approach and examples of how these theoretical components fit together in implementation by our teacher colleagues. Our goal is to support your ability to apply these ideas in your own context.

A useful distinction for thinking about how to learn challenging material is analytic versus synthetic thinking. In *analytic* thinking, learners separate a complex system into parts and examine each part to understand how it functions in isolation. This allows learners to begin to understand what each part contributes to the larger system. Analysis is particularly useful to simplify learning when initially approaching a complex system, but complex systems function in important ways that go beyond the sum of their parts. How a part acts in isolation often differs from how it acts in a complex system. Focusing on *synthesis* requires looking at the larger system and how the parts interact to influence one another. For this reason, analytic thinking is a good starting place, but synthetic or systems-level thinking is required to understand complex integrative systems. Teaching and learning in real classroom contexts certainly qualifies as a complex system!

In this chapter, we separate the core components of our larger model and focus on understanding each individually. Our goal is to use this analytic approach to build foundational understanding, and in later chapters focus on synthesis and real-world applications to help you understand more deeply so that you can apply what you learn to your teaching and learning practice. The core components we focus on in this chapter include: (a) the design process with mathematical modeling and waves of divergent and convergent thinking and (b) the integration of mathematics, science, and literacy content standards and practices. As you read about each of these separately, keep in mind that these parts are integrated into a larger system, and understanding how synthesis in that system informs teaching and learning practices will be the focus of the remainder of this book.

A VISUAL MODEL OF OUR APPROACH

Figure 2.1 is a visual model of how the design process with mathematical modeling connects to content standards in science, mathematics, and literacy. You will recognize this as a somewhat expanded version of Figure 1.2. Before reading further, we suggest you take a few moments to think about what you know about each of the pieces in this model, how you currently understand each individually, and how you think they fit together. Maybe even jot down ideas or questions that arise for you as you look at Figure 2.1. Then continue reading with these ideas in mind and return to them as you finish the chapter. As you likely know, research shows that reflection and active reading such as this helps connect what learners read with what they already know (National Academies of Sciences,

Figure 2.1. Expanded Model of Our Approach

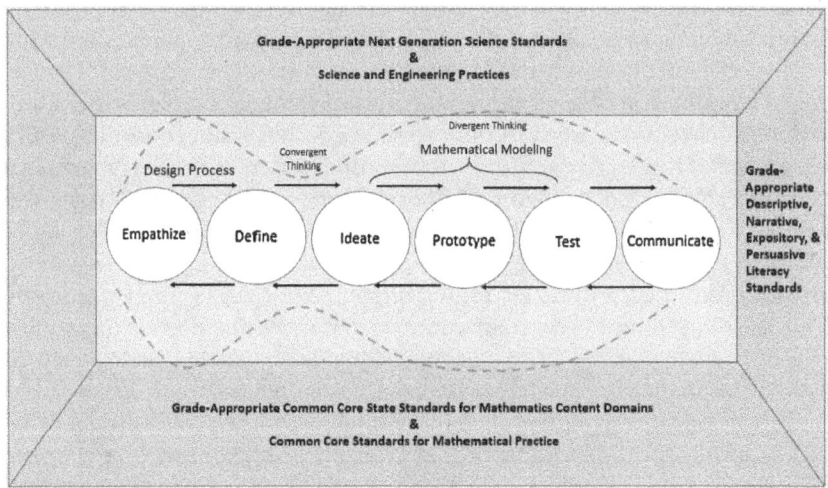

Engineering, and Medicine, 2018); engaging in this approach will help you connect what we share in this book to reflection on your teaching practices. At the center of Figure 2.1 is the design process utilizing mathematical modeling and embedded in waves of divergent and convergent thinking. Surrounding the design process on three sides are content standards and practices from science, mathematics, and literacy. We focus the remainder of this chapter on understanding each of these core components of our larger model. While the power of this model comes through the integration and application of these core components in teaching and learning practice, understanding each separately will provide you with a strong foundation to build on throughout the rest of this book.

One final note about how we discuss teachers and their students: We have found through our professional development that teachers learn this material best when placed in the position of learners themselves. That experience as learners gives teachers important insights into how best to support their students' engagement in design-based learning. For that reason, in the sections that follow when we say "you," we are referring to any learner engaged in design-based learning, whether that is yourself (teacher as learner) or your students.

THE DESIGN PROCESS: DYNAMIC, ITERATIVE WAVES OF DIVERGENT AND CONVERGENT THINKING

At the very center of Figure 2.1 you see the stages of the design process represented with a series of circles, each with an action term inside of it: Empathize, Define, Ideate, Prototype, Test, and Communicate (see Figure 2.2). Before explaining what each of those actions entails, we want to warn you against a common misconception about the design process.

It may be tempting to think of the design process stages as a series of steps to follow, one after the other, to accomplish a "complete" design. As tempting as that may be, it is a common but incorrect way of understanding the design process. Rather, the arrows pointing in different directions

Figure 2.2. Design Process Stages

- Empathize: Focus on what users need.
- Define: Identify problems your design should solve.
- Ideate: Develop multiple potential solutions and parts of solutions.
- Prototype: Build initial scale model(s).
- Test: Try out scale model(s); record and analyze performance data.
- Communicate: Describe design process, findings, and suggestions for next steps.

in Figure 2.1 are meant to highlight the dynamic and iterative nature of the design process. The *dynamic* part means that each stage focuses on action and reflection, while the *iterative* part means that the action and reflection may or may not result in moving to a different stage at any point. Instead, you should think of each circle as a category of action to focus on and then reflect about whether what you have learned positions you to continue more activity in the current stage, or to move to another stage either to the left or to the right. Cycling to the left to redefine, redesign, and retest are critically important parts of the design process.

The wavy lines above and below the circles in the design process are meant to represent how divergent and convergent thinking are emphasized differently across the various stages. *Divergent thinking* focuses on creativity and diversity, generating as many different answers as possible. *Convergent thinking* focuses on narrowing choices, finding the "best" possible answer in a given context. When mathematical modeling is used as a tool for convergent thinking in this context, learners come to see the usefulness of mathematics more concretely. We think of these two modes of thinking as overlapping waves that ebb and flow across the stages, where diversity and multiple potential approaches or solutions are foregrounded in some stages, while other stages are focused on selecting among competing alternatives and converging on the "best" solution so far.

With an understanding of what it means for the design process to be dynamic and iterative, we are ready to consider the kinds of action, reflection, and divergent/convergent thinking involved in each of the six design process stages. The focus of each stage is briefly described in Figure 2.2, and here we discuss why that focus is important, considerations for how to accomplish that focus, and implications for next steps in each stage. If you would like to explore further, we recommend resources available from the Institute of Design at Stanford as a good starting place (https://dschool .stanford.edu/resources).

Empathize

This stage is the hallmark of what is called *human-centered design*. The focus is on understanding the person or people who will use the process or product created through the design process. Who are they? What are their needs? What will they be trying to accomplish? What about them or their environment could affect their use of what you design?

The best way for designers to understand their users is usually to talk to them and observe them doing things related to the problems the design should help solve. In some cases, conducting formal interviews and observations makes sense, but more often casual conversations with potential users (who may even be members of your design team) are a good start.

Divergent thinking is important here, as designers want to generate as many different potential understandings of their users as possible. People are diverse, and you want to capture that diversity. At the same time, convergent thinking is important because you want your understanding of user needs to converge on those aspects that are most closely related to the design challenge you are trying to address.

As you begin to consider whether it is time to move to another stage, it is important to get your ideas out on paper, a whiteboard, or some other place that you can refer to and talk about them with others. The main question to be asking yourself at this point is, "Do I understand my users well enough that whatever I design is likely to be something that will actually help them and that they will want to use?"

Define

This stage is focused on clarifying the problem(s) your design should solve. As such, the information gained in the Empathize stage is an important part to consider. It is also important to consider what a "good" design should entail—that is, what assessment criteria can be used to determine if one design is better than another?

The goal is to describe a challenge or problem that is meaningful and important to real people, and that designers can conceptualize potential solutions to that can then be compared or tested to determine which potential solution works best. You need to consider the user, the context in which whatever you design will be used, and assessment criteria for design "success."

There are some elements of divergent thinking in coming up with multiple potential design assessment criteria, but convergent thinking is at the fore. The result should be a paragraph or so detailing the specific problem(s) you are trying to solve, the constraints or limitations that good solutions must satisfy, and assessment criteria for comparing one solution against another.

Such a clear design definition prepares you to move into the Ideate stage. However, it is not uncommon to realize that you need more information and return to the Empathize stage. It is also possible that the Define stage so clearly implicates a specific design that you choose to jump ahead to the Prototype stage; this sometimes occurs, for example, when there are a lot of design constraints. However, it is important not to let biases in your initial ideas drive your design, and the Ideate stage is important for bringing out a diversity of potential solution paths.

Ideate

This stage is focused on shifting from defining the problem to coming up with potential solutions. It is driven by divergent thinking and focused on generating as many possible solutions and approaches to the design as possible.

This is commonly called brainstorming and should not focus initially on whether any solution is complete or good or feasible. Rather, the focus should be on getting as many different ideas out as possible. The more diverse your design team, the better. Everyone's minds are needed. All ideas are welcome. Off-the-wall, out-of-the-box thinking is what you want because that is often where truly innovative and transformational solutions come from.

Ideas often come fast and furious in this stage, so be sure to capture them in writing as they do. Think of the ideas generated as the ocean from which you can pull what you decide are the most promising possibilities to try out. The wider and deeper that ocean is, the more options you have, and the more likely great solutions are to be in there somewhere. Crazy, far-fetched ideas are important, and it is critical to solicit weird solutions. You never know where a truly revolutionary solution will come from.

As divergent thinking in the Ideate stage proceeds, you eventually need to begin applying convergent thinking to determine which ideas to move forward into the Prototype stage. You will know you are reaching this point when no new ideas for potential solutions or approaches are coming forward anymore.

A useful approach at this point is to get all of the ideas generated so far into a form where they can easily be moved around and grouped. We suggest using sticky notes or index cards with one idea written on each note or card, and then grouping the ideas by similarity, part of the solution focused on, or whatever else seems to make sense to the design team. This will allow you to see common themes and think about which ideas or groups of ideas can be the basis of a prototype. As you consider what ideas to prototype, remember to use the results of the Empathize and Define stages to think about which ideas are most likely to best meet the design challenge and user needs. As we will discuss more later, using mathematical modeling at this point to predict and select which approaches may be more effective or efficient is a core component of our approach.

Prototype

This stage involves building inexpensive scale models of design solutions and parts of design solutions. You should build at least two or three prototypes and reflect on what you are learning through this process. While building more than one prototype may at first seem wasteful if you have started with what you think is the best set of ideas for your first prototype, it often turns out that important new ideas come from combining two or more possible solutions. Prototyping multiple solution paths and reflecting on what you learn as you build and then test them gives you insights that would not otherwise come to the fore.

It would seem that this is mostly convergent thinking, as you are focusing in on a solution to test. However, in practice it becomes more complex than that. The goal is to move these prototypes to the Test phase, but you will almost always find that the very process of building prototypes requires cycling back through the Ideate stage. As you build something, you often find that it does not come together as you thought it would. The building of the prototype itself may present unexpected challenges; finding potential solutions to these challenges requires divergent thinking. Sometimes these challenges make you realize there is something about the problem definition that needs revisiting, so you cycle back to the Define or even Empathize stage to refine your understanding.

Once you successfully build a prototype to test at least part of your design definition, you are ready to test that prototype. Cycling among Ideate, Prototype, and Test phases with multiple combinations of ideas and prototypes should be the goal.

Test

This stage is focused on testing one or more of the prototypes you have built, collecting performance data on the assessment criteria identified in the Define stage, and comparing prototypes. This gives you critical information for redesigning and building either additional prototypes or a full-scale model.

Testing is primarily about convergent thinking, as you are using the information you gain to narrow down design choices. However, it is not unusual for testing to reveal substantial problems that require you to return to earlier stages where you once again need to focus on divergent thinking as you go back to the drawing board, so to speak.

Communicate

This stage is focused on consolidating what you have learned through the design process so far, clearly communicating that information to get feedback from others, and considering what the next steps should be. Most of the time, next steps involve returning to other stages and continuing the dynamic iterative process to (re)design solutions.

After multiple iterations, you may have developed a good solution and evidence that supports its utility and efficiency. In that case, you may be using the Communicate stage to try to convince decision-makers to put your solution to practice at a larger scale.

INTEGRATING MATHEMATICAL MODELING, SCIENTIFIC CHALLENGES, AND LITERACY PRACTICES

In some ways, the argument for integrating across content areas is straight-forward: It is a set of opportunities that should not be wasted. Teaching mathematics as numerical tools devoid of real-world scientific content can make it harder to learn and harder to motivate learners. Teaching science without drawing on the tools of mathematics for things such as measurement, data representation and analysis, and addressing precision and variability is an opportunity wasted. Helping students come to understand mathematics and science content while developing evidence of the usefulness of what they learn and build, but then not teaching them to communicate to persuade others and reflect on what they are learning, is another wasted opportunity. Wasting such opportunities is a great disservice to today's youth as they prepare for full participation in 21st-century society.

Of course, it is important to recognize that this should not all fall on teachers alone. Traditional teacher preparation, teacher certification, school structures, and educational assessment and accountability do not tend to provide robust support for cross-curricular integration or the development of teacher expertise outside their primary content area. It is difficult for science teachers to acquire deep mathematics or literacy content expertise. The same is true of mathematics or English language arts teachers developing expertise outside their specific content area.

One of the most effective approaches we have found in our work has been cross-content teams or networks within or across schools. In fact, as single-subject (mathematics or science) teachers begin implementing our approach, they often begin to grow cross-disciplinary networks of their own. Teachers with different content-area backgrounds can support one another through such networks, and, where possible, curriculum specialists, teacher educators, and university content-area specialists can help as well. Design projects can be of vastly differing scales, and collaborations can begin on relatively short and simple design activities. Education policymakers, teacher educators, and education administrators at every level need to work to make space for such networks and to support teachers as they strive to meet the call for content integration.

Returning to the argument for integrating across content areas, we will be more explicit for those familiar with national standards documents, but please recognize that later chapters include specific examples of how to accomplish this integration. In our approach, mathematical modeling is a key tool aiding content integration. Mathematical modeling has been increasingly emphasized in standards documents over the years, as increased focus has been placed on real-world applications of mathematics (Cirillo et al., 2016). The Common Core State Standards for Mathematics (CCSS-M) includes "model with mathematics" as one of the Standards

Figure 2.3. Themes in Mathematical Modeling Definitions

- Mathematizing real-world, authentic problems that may not initially appear "mathematical" and may not have a single solution.
- A process that can be used to explain, predict, or analyze real-world events.
- An iterative process in which learners use their mathematical knowledge and creativity to make choices, assumptions, and decisions to find the most appropriate solution.

for Mathematical Practice (SMP) for K–12. "Modeling" also appears as a domain in high school standards. While there are many definitions of mathematical modeling, common themes of these are shown in Figure 2.3.

In our approach, mathematical modeling is a key tool during the Ideate-to-Prototype stages in the design process as learners determine significant variables, determine how to quantify them, create mathematical models to represent the situation, and then, importantly, analyze to choose among alternative designs. Concurrently, mathematics content domains (e.g., ratios and proportional relationships, statistics and probability) and standards for mathematical practice (e.g., making sense of problems and persevering in solving them, reasoning abstractly and quantitatively, choosing appropriate tools) are integrated with science and engineering content standards and practices from the Next Generation Science Standards (e.g., asking questions and defining problems, using mathematics and computational thinking), as well as cross-cutting concepts focused on systems and system models. Design projects provide extensive opportunities to engage in practices common to mathematics and science standards documents: defining problems, constructing explanations, developing models, and attending to precision, as well as integrating literacy standards and practices for communication and reflection.

The idea is to motivate students with engaging problems related to their lives and help them use the design process to develop solutions to those problems. Along the way, teachers identify specific grade-appropriate science, mathematics, and literacy content standards and objectives (CSOs) that connect to design challenges students are working on. And, as we will describe further in Chapter 3, teachers also apply the design process to designing and redesigning their lessons to solve the "problem" of how best to support learning relevant content.

Single-subject (mathematics or science) teachers can do this in their classrooms, and even though one content area may be the dominant focus, every design lesson should have CSOs from mathematics, science, and literacy domains. Mathematics CSOs are needed as conceptual tools to structure information, measure, collect data, analyze data, and predict which solutions may be most helpful. Science CSOs are needed to understand the context of design problems and how potential solutions are likely to function. The Prototype and Test stages of the design process let learners experience mathematics and science in action so that they develop deeper understanding of that content. Literacy CSOs come into play as students describe and reflect on what they are learning at each stage of the design process, write explanations and reflections on relevant science and mathematics content, and write to persuade others that their designs are valuable or in some cases even worthy of scaling up with real-world investment.

We believe mathematical modeling should have a more prominent place in middle grade science and mathematics. As discussed in the Introduction, we distinguish our design-based approach from more commonly used project-based and problem-based approaches. The reason we revisit this distinction here is that project-based and problem-based learning have been fairly well described and often implemented in many science and some mathematics classrooms for a long time. However, our review of the literature on project-, problem-, and design-based learning, as well as our experiences working with teachers, have led us to understand that too often project-, problem-, and even some design-based learning projects have very little meaningful mathematics content.

While that may not seem like much of a problem for a science classroom, it turns out that not bringing in relevant mathematics makes it harder to learn science. Consider, for instance, trying to teach and learn about force and momentum without measuring speed or mass. You could address the movement of physical objects in general terms, but integrating mathematics into that science content would result in a much more precise understanding. And how better to reinforce learning than to predict what should happen and then try it out to see if your predictions are accurate? Science teachers who do not bring in relevant mathematics, in our observations, tend to shy away from important understandings of science that are difficult to describe in nonmathematical terms. The reverse is also true. Mathematics taught devoid of real-world applications that science content can provide is much more difficult to learn and less engaging for many students. Mathematics teachers who do not bring in relevant science contexts, in our observations, tend to have difficulty convincing students of the usefulness of the mathematics they want them to learn.

Another example from one of our teacher colleagues will help you see how the core components of our approach fit together. Consider Ms. Vaughan's 8th-grade science classroom. She wanted to increase student engagement and provide opportunities for deeper learning in understanding 3D models of molecules. In previous years, Ms. Vaughan had used cocktail sticks and candy to have the students make 3D models, but despite the fun nature of the activity, she felt that her students were not acquiring the deep understanding of content she wanted them to have.

Figure 2.4. Standards in the 3D Molecules Activity

CCSS-M Mathematics Content and Practice Standards

- 6.RP.2: Understand the concept of a unit rate a/b associated with a ratio $a:b$ with $b \neq 0$, and use rate language in the context of a ratio relationship.
- MP1: Make sense of problems and persevere in solving them.
- MP3: Construct viable arguments and critique the reasoning of others.

NGSS Science Standards

- MS-PS1-1: Develop models to describe the atomic composition of simple molecules and extended structures.
- MS-ETS1-2: Evaluate competing design solutions using a systematic process to determine how well they meet the criteria and constraints of the problem.

CCSS-ELA Standards

- W.8.4: Produce clear and coherent writing in which the development, organization, and style are appropriate to task, purpose, and audience.

Ms. Vaughan worked on adapting the lesson to incorporate design-based learning and focused on multiple standards using an activity for students to design and build kits for making 3D models of molecules (see Figure 2.4). Ms. Vaughan framed the question for this year's students as follows:

> Sunnydale Middle School's 8th-grade science class needs molecule kits for students to practice building molecular models. The school cannot afford to purchase the actual kits, and the cheapest kit online was $29.99. Your task is to create an affordable molecule kit for the school so they can practice building molecules. Each kit must include: (a) some everyday material to use for the bond (30 bonds), (b) some everyday material to use for each element (carbon, hydrogen, nitrogen, oxygen), (c) calculation of cost for each material used and total amount, including whether the materials will be reusable or purchased again each year, (d) an information key to show what materials represent each element, and (e) an introduction to building molecules with basic information. After you create your kit, you will make a brochure to inform Sunnydale about your kit and why they should purchase the more cost-efficient kit. Reflect on how you developed your kit: How did you decide which elements and bonds to include, and how many were needed? What types of molecules could you make with your kit, and where can they be found? Are you able to make any connections with things you have studied in science or math in the past?

The modified design-based lesson plan provided many opportunities for students to ask one another reflective questions about elements and bonds. The brochures also led to student questions. Ms. Vaughan was particularly pleased with how students explored science content as they reflected on similarities and differences across molecular models.

This science-forward design-based learning example highlights that many existing lesson plans can be modified to incorporate design-based learning, provide students with more ways to engage with the material, and integrate literacy in meaningful ways. We discuss how to modify existing activities and textbook problems to create engaging design-based learning activities in Chapter 12.

CHAPTER 3

Affordances and Supports

You now have a developing understanding of the core components of our approach: design-based learning, mathematical modeling, and math-science-literacy content integration. You have read about each of these pieces separately as well as examples of real teachers putting them together in their classrooms. In this chapter, we expand on these core components with two additional categories: affordances and supports. *Affordances* are teaching challenges that our approach is particularly helpful for addressing. We focus on fostering equity in the classroom and scaffolding productive struggle as key affordances. *Supports* are considerations that help this approach be most effective. We focus on complex instruction and reflective teaching as key supports in our approach.

We will continue to write about you (our reader) both as a teacher and as a learner, so that the learners we refer to are sometimes your students and sometimes yourself. This is a purposeful positioning based on the idea that reflective teaching and teachers-as-learners are important supports for our approach. In that vein, consider the way we are gradually building up the visual model of our approach. In Chapter 1, we presented the most simplified version and then expanded on that with more detail in Chapter 2. In this chapter, we present the full model, including affordances and supports (see Figure 3.1). Notice that this gradual building-up of a conceptual structure and connection of real-world examples to it is a useful way of thinking about what teachers hope happens for their students as they acquire ever more sophisticated understandings of mathematics, science, and literacy through design-based learning.

Now that you understand the core components of our approach, you are in a position to ask two critical questions that we address in this chapter: What is it good for? And what does it take to make it work? Answers to "What is it good for" questions can be called affordances, which are what something allows you to do, such as how a shovel can allow you to dig a hole. The most obvious of affordances for our approach will be the focus of Chapter 4, as we describe the evidence for why we believe this approach works to effectively bring about deep and integrated student and teacher learning. In this chapter, we begin with a discussion of two affordances

Figure 3.1. Full Model of Our Approach

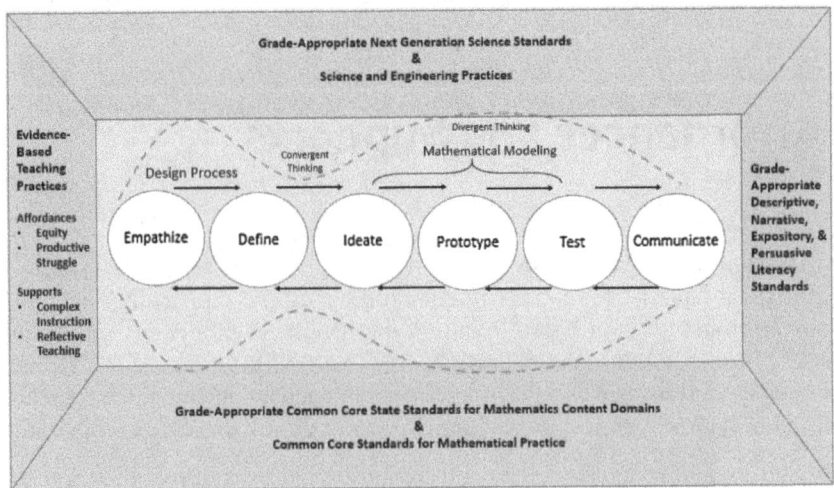

and why they are important. We follow that with concrete discussion of supports that enable those affordances.

The affordances we focus on in this chapter include supporting productive struggle by learners and promoting equity in the classroom. Complex instruction is explicitly designed to support equity in classrooms, and reflective teaching is a critical support for all aspects of our approach. As will be described in Chapter 4, each of these affordances and supports is an *evidence-based teaching practice*. The key idea here is that teaching practices should be grounded in evidence from research on teaching and learning as well as from teachers' experience in the classroom and knowledge of the specific learners they serve.

SUPPORTING PRODUCTIVE STRUGGLE BY LEARNERS

Productive struggle is the process of effortful learning that occurs when learners persevere through difficult problems. Persevering through productive struggle and learning through failure are two ways of describing this key affordance that has emerged as central in our work with teachers and their work with students. Educational research has long examined and highlighted the benefits of engaging in productive struggle or even "productive failure" in the learning process (Kapur, 2008, 2010). Productive struggle is a key feature in learning that is conceptual, robust, and transferable. In mathematics, where it has been studied most, engaging students in productive struggle has been found to support students' conceptual understanding, and as a result, it is one of the "ambitious teaching practices" encouraged for creating high-quality, rigorous, and equitable mathematics learning for all students (Hiebert & Grouws, 2007; NCTM, 2014).

Yet, like design-based learning, engaging students in productive struggle while learning can be challenging. Most adults (including teachers) did not learn in this way, so it may not be a "typical" approach to teaching and learning for them. Further, some teachers may be uncomfortable with their students' struggle and, instead, want to remove struggle so that students "succeed" at a task. Part of that difficulty comes from the dominant cultural view of mathematics and science as only for "some people" or as a static body of knowledge that must be learned rather than created. The idea of engaging in struggle while learning may counter a traditional view of mathematics and science as something you either "get" or not. In that perspective, struggle would only be for those who do not get it. Such beliefs can lead to instructional practices that seek to remove struggle from students rather than leverage it for learning (Stigler & Hiebert, 2004).

Productive struggle hinges on instructional tasks that investigate content and create knowledge in meaningful ways. Therefore, instructional approaches that engage students in productive struggle also challenge existing notions of what it means to do mathematics and science; this can be uncomfortable for parents, other teachers, administrators, and some students. Successfully supporting productive struggle requires a good bit of understanding of both the content being taught and how students might respond to that content, including any challenging points or common incomplete understandings. It is difficult to develop appropriately challenging tasks and decide how much autonomy versus support, sometimes called *scaffolding*, to provide students with so that they are genuinely challenged but not overwhelmed. Yet the challenges do not outweigh the potential gains in understanding and positive identity that can benefit students.

Making these challenges and connections explicit supports teachers' developing understanding of how to effectively engage their students in productive struggle, and the design process is particularly supportive of reframing failure as productive. As noted above, we define productive struggle as effortful learning while persevering through difficult problems. This applies to the act of teaching as well. Teaching certainly requires persevering through a diversity of difficult problems related to how best to help a given group of students. This provides a clear connection to the reflective teaching practices described below, one of several integrative connections among affordances and supports for our approach.

PROMOTING EQUITY IN THE CLASSROOM

A second important affordance is how our approach can *promote equity* in the classroom. At the most basic level, equity is fairness. One way of thinking about this is that if we achieve equity, then we will not be able to predict differences in student outcomes based solely on things like race, class, or gender (NCTM, 2014). Rochelle Gutiérrez (2012) developed a framework for thinking about equity for mathematics teaching and learning that is useful for our work (see Figure 3.2). This framework includes two axes and four dimensions of equity: access and achievement (on the dominant axis), and identity and power (on the critical axis). Understanding these four dimensions broadens what equity means and allows more effective

Figure 3.2. Rochelle Gutiérrez's (2012) Equity Framework

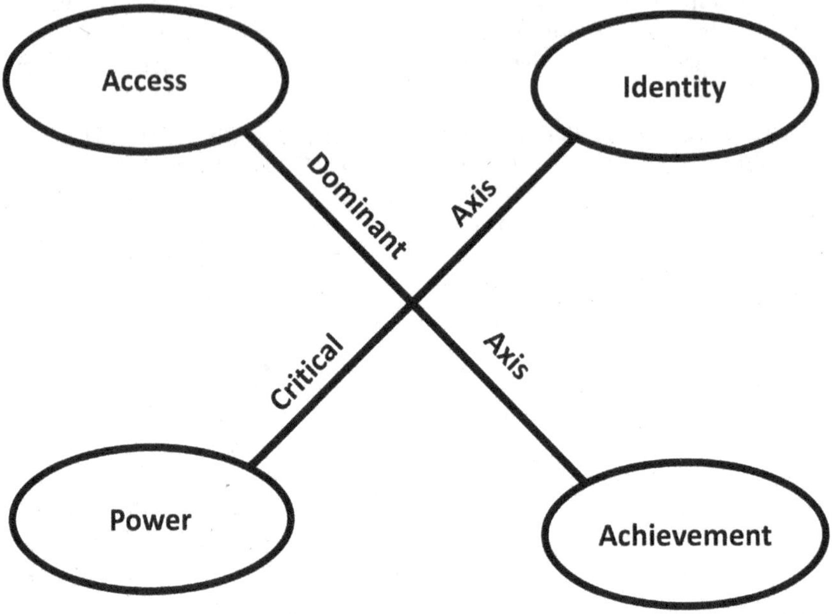

Access: Tangible resources allowing high quality opportunities to learn.

Achievement: Tangible results demonstrating learning.

Identity: Students can see themselves and understand others through the curriculum.

Power: Students have voice and agency in the classroom and beyond.

Source: Adapted from Gutiérrez (2009).

action to promote equity in classrooms, with the potential to improve outcomes for individual students and transform society at large.

- *Access equity* is about fair distribution of resources. Do all students get high-quality teachers, challenging curricula, needed supplies, and support to participate in learning in and out of classrooms? Is the classroom environment one that encourages and welcomes all students' contributions?
- *Achievement equity* is about fair distribution of learning outcomes. Do all students take the courses and pass the tests that act as gatekeepers for education and career paths?
- *Identity equity* is about fair representation. Do all students see themselves represented in the curriculum; do they see themselves as able to participate in the content; are they able to leverage their lives and experiences for learning both about themselves and about other groups who may be less familiar to them?
- *Power equity* is about fairness in who contributes and what matters. Are all student voices valued, and are students able to use what they learn to critique society? Are they supported in using learning to improve their worlds?

Access and achievement equity are driven by dominant cultural conceptions of what is valued. As such, while individual teachers can address some aspects of these dimensions, they also rely on the influence of broader changes in social policy. In our perspective, identity and power equity, the critical dimensions, are those that teachers can more directly influence in their classroom. These dimensions focus on challenging the status quo of who can do mathematics and science, and what that "doing" looks like. They connect with the human side of these disciplines: how they can be used to address issues and challenges that are meaningful to students. This directly aligns with the human-centered design approach described in Chapter 2. Focusing on these dimensions drew us to the supports we discuss next.

As we developed our approach over the years, two evidence-based teaching practice supports that have been helpful for us and our teacher colleagues are the use of complex instruction and reflective teaching practice. These evidence-based teaching practices support a critically important focus on equity in the classroom and engaging learners in productive struggle.

COMPLEX INSTRUCTION

Complex instruction is an evidence-based teaching practice developed at Stanford University that focuses on improving equity in cooperative learning classrooms (https://complexinstruction.stanford.edu/about). There is substantial evidence that complex instruction results in positive student learning outcomes and is particularly effective in diverse or heterogeneous classrooms. We were drawn to two aspects of complex instruction, in particular: the use of group-worthy tasks and addressing issues of status in the classroom.

Group-Worthy Tasks

There is strong research evidence showing improved academic and socioemotional outcomes from engaging students in effective group work and collaborative learning (Watanabe, 2012). The foundation of effective group work is what Rachel Lotan (2003, 2006) called *group-worthy tasks*. These are tasks around which every member of a diverse group of students can usefully contribute and learn; thus, they are critical tools for improving student learning and equity. There are five components that must be present for effective group-worthy tasks (see Figure 3.3).

Group-worthy tasks are not clearly defined, with straightforward paths to one correct answer. Rather, they are open-ended tasks that require complex problem solving and have multiple entry points and, in some cases, multiple possible "correct" answers. Group-worthy tasks require a diversity of skills, and learners can contribute in many different ways so that all learners are engaged. Group-worthy tasks focus on important content in specific disciplines, although one task may draw on content across multiple disciplines. Group-worthy tasks include both individual group member accountability for their contributions and accountability of the whole group for what the group produces. Group-worthy tasks have clear assessment criteria for evaluating what the group produces.

We will share examples in later chapters of group-worthy tasks embedded in our integrated approach. To give you a sense of what we are talking about now, we encourage you to search for Dan Meyer through the Internet browser of your choice and watch his "Math Class Needs a Makeover" talk (Meyer, 2013). One of the things he does is take a standard math textbook word problem and turn it into what could be used as a basis for a group-worthy task. One example problem includes an empty

---

Figure 3.3. Five Components of Group-Worthy Tasks

- Open-ended
- Multiple entry points
- Meaningful content
- Group and individual accountability
- Clear assessment criteria

---

water tank where the height and width measurements are given and a set of substeps to get from those measurements to the correct answer are also given. This straightforward and well-defined problem is not a group-worthy task. What Dan Meyer does is strip away all the measurements, remove the substeps, show students a video of someone actually filling a water tank with a hose, and then engage students in conversation about how they could use math to figure out how long it is going to take for that process to finish. Students have to decide what is worth measuring, how to measure it, and what to do with the measurements to determine the accuracy of their initial guesses about how long it would take. Add to that clear assessment criteria for group and individual accountability, and there you have a group-worthy task.

Status

Complex instruction includes the use of group-worthy tasks while also attending to and addressing "status." *Status* is about how students see themselves and others. Status determines who believes they can enter academic conversations and contribute in group-learning contexts. Anyone who has been in a classroom has noticed that some students participate regularly, while others participate little or not at all. Status drives who participates more, and therefore who learns more. Status is socially determined and is influenced by race, class, and gender, but also by individual characteristics like reading ability, past math/science learning, and peer social influences.

A key insight of complex instruction is to recognize the influence of status and how to take steps to more equitably distribute status in the classroom. Complex instruction includes focus on physical structure such as arranging desks so that all group members face the task and the work remains in the center, so all group members can equally reach learning materials and contribute to solving problems. Complex instruction includes explicit focus on multiple abilities, not just traditional math or science skill, and explicit discussion about how everyone can and must contribute to cooperative learning using their unique abilities. Complex instruction includes seizing on opportunities to recognize students' contributions, particularly those who may not often share, and *assigning status* to them. This process publicly (within the group and to the whole class) and explicitly brings forward and highlights the contributions of all students to the group's success, including those who may be perceived as holding "lower" status in the classroom community. The goal of these efforts is to create heterogeneous cooperative learning groups where everyone contributes and learning leverages all group members' abilities.

Challenges addressed with the design process are rich foundations for complex instruction. Mr. Shaker's shoebox design challenge was a good example of this. Remember how two of Mr. Shaker's students whom he previously had difficulty engaging developed an innovative triangular prism shoebox design. Mr. Shaker publicly shared their success with the class, influencing their status and self-efficacy as contributors to mathematical and scientific thinking. Complex instruction implemented in ways that encourage and recognize contributions by all members of the group both promotes equity and provides opportunities to engage and support students in productive struggle while learning. .

REFLECTIVE TEACHING PRACTICE

The final evidence-based teaching practice support we focus on in this chapter connects and runs through all the other components of our approach. Remember how we emphasized reflection for you actively reading this book and how reflection featured prominently in each stage of the design process in Chapter 2. Reflection in general is a form of *metacognition*, which means thinking about your own thinking, and research shows that it can be taught and benefits learners who practice it (National Academies of Sciences, Engineering, and Medicine, 2018).

In this section, we focus more specifically on reflection by teachers about their teaching practice and how that reflection relates to their students' learning. In other words, we focus on reflective teaching as an evidence-based teaching practice. We also introduce the idea of applying the design process to the "problem" of teaching and learning, using it as a tool for reflective teaching.

Reflective teaching is thinking about and critically analyzing evidence related to the impact of your teaching on learners in order to improve your teaching practice. Our understanding of reflective practice is largely founded on the influential work of John Dewey and Donald Schön. Dewey (1933) emphasized the active nature of reflection beyond casually thinking about something and argued that it requires, among other things, empathy, curiosity, self-awareness, communication skills, patience, risk taking, and actively seeking feedback. Dewey described five phases of reflective thought that you may notice overlap considerably with the design process (see Figure 3.4).

Schön (1983, 1987) expanded on Dewey's work, including the idea that expert practitioners reflect in at least two ways: reflection-in-action and reflection-on-action. *Reflection-in-action* occurs in the moment while you are teaching, and much of what guides it is below the level of consciousness unless you are intentionally careful to focus on it. Through education and experience, teachers develop a wealth of knowledge about how to adjust their teaching in the moment based on what is happening in the

Figure 3.4. John Dewey's Phases of Reflective Thought (and Related Design Phases)

- Suggestion (Empathize): A challenging experience prompts initial intuitive solutions to be considered.
- Intellectualization (Define): The challenging experience is clarified into a problem to be solved.
- Leading Ideas (Ideate): Possible solutions lead observation and collection of additional relevant information.
- Reasoning (Ideate, Prototype): Leading ideas and additional information collected are combined and recombined to elaborate a potential solution.
- Testing (Test): The elaborated and refined potential solution is tested either in action or through imagination.

classroom and their own beliefs and theories about what works in teaching and learning. Teachers use this implicit knowledge with reflection-in-action to quickly take in feedback from students, consider that feedback in the context of instructional goals and beliefs about how students learn, and adjust their teaching practices in real time without too much conscious thought because there simply is not time. While much reflection-in-action is implicit, or below the level of consciousness, teachers can learn to focus on their implicit knowledge and reasoning, making it explicit so that it can even better inform their teaching practices.

Reflection-on-action occurs after teaching is done and teachers have time to intentionally consider what happened, how they felt about it, and evidence related to the impact of their teaching. *Student assessment* is an important information source for reflection-on-action and should take many forms: tests and quizzes (standardized, textbook, teacher-made), writing assignments (descriptive, narrative, expository, persuasive), and authentic tasks (making and doing with rubrics applied to evaluate performance). Reflection-on-action is greatly facilitated by systematically gathering evidence related to instructional practices, beliefs, and impacts. Capturing what happens in the classroom, your thoughts and beliefs about what is going on, and evidence related to student learning is critical to supporting effective reflection-on-action.

An important component of reflection-on-action should be a focus on equity, and we will discuss connections among reflection and other evidence-based teaching practices focused on equity, such as complex instruction, with specific examples of how they fit together, in later chapters. For now, it may be enough to say that considering who your students are and how they interact with you and with their peers in the classroom is a fruitful area for reflection regarding how your instructional practices impact equity in your classroom.

LEVERAGING THE DESIGN PROCESS FOR
REFLECTIVE TEACHING

You now understand something about the core components, affordances, and supports of our approach and no doubt see many connections and interactions among them. Looking back at Figure 3.1, notice again that the design process is front and center. The content that students need to learn and utilize in developing design solutions surrounds the design process on three sides, while evidence-based teaching practices initiate and guide learning. Evidence-based teaching practices include affordances and supports such as complex instruction and reflective teaching in order to help students struggle productively and develop strong conceptions of themselves as learners with the status to make valued contributions to meaningful tasks in the classroom. All the core components, affordances, and supports in our approach can be taught and improved through practice; teachers can learn and continuously improve their use. Reflective teachers continuously examine and strive to improve their instruction and its impact on learners.

We want to point out one last conceptual tool as we wrap up this chapter: leveraging the design process for reflective teaching (see Figure 3.5). We believe this is an incredibly important and valuable way of thinking that provides much of the power of our approach. When you apply the design process to reflective teaching,

- you get to know your learners (Empathize);
- you identify knowledge and skill growth opportunities through student assessment and connect those to content standards and objectives (Define);
- you develop multiple instructional activities and practices to help close knowledge and skill gaps (Ideate);
- you create lessons by combining and recombining instructional activities and practices (Prototype);

Figure 3.5. The Design Process Applied to Reflective Teaching

- Empathize: Get to know your students and what they need to learn.
- Define: Connect what your students need to learn to specific content standards and objectives.
- Ideate: Brainstorm multiple instructional activities and practices.
- Prototype: Develop activities and lessons.
- Test: Teach lessons, assess student learning, and practice reflection-in-action.
- Communicate: Practice reflection-on-action and share what you are learning.

- you try out your lessons (Test) in ways that help reflection-in-action become explicit and provide rich information for reflection-on-action to inform future teaching; and
- you share what you learn with colleagues and students (Communicate).

Throughout reflective teaching practices that leverage the design process, it remains important to remember that the design process is dynamic and iterative. It is not completed with a single pass through the phases, and it cannot be successfully applied without critical reflection.

CHAPTER 4

Why Do We Think This Works?

Now that you have an understanding of the core components, affordances, and supports for our approach, we focus in this chapter on why we think this approach works so well to support deep and integrated student and teacher learning. Our goal is to share the evidence supporting our approach and why we believe it can be transformative for your teaching and learning. To do that, we draw on educational research published in academic journals and books, as well as evidence we developed over more than a decade working with teacher colleagues engaged in a series of professional development projects.

The first parts of this chapter are organized around evidence-based teaching practices, including complex instruction, productive struggle, and reflective teaching. Next, we address the evidence base for our core components: content integration, the design process, and mathematical modeling. We conclude the chapter with a discussion of evidence we have developed related to the value of integrating the full model. This last section serves as a bridge to Chapter 5 and the rest of this book, where we focus on putting it all together to personify effective teaching practices.

EVIDENCE-BASED TEACHING PRACTICES

The beginning of the 21st century saw a strong push for evidence-based teaching practices through federal and state educational policies (e.g., No Child Left Behind). Unfortunately, only specific kinds of evidence counted. Only things that worked to increase standardized test scores on average across many different classroom and school contexts were considered "evidence-based." While this produced some useful practices grounded in learning theory and demonstrating broad applicability, educational researchers increasingly realized what many classroom teachers might have told them all along: that context matters on a local and classroom level, and student learning involves much more than standardized test scores.

Every community, school, classroom, teacher, and student is different, and those differences interact to create classroom cultures and contexts that impact what works. What works for some, or even most, learners will not work for others. Students' (and teachers') views of themselves as learners and their place in the classroom community matter. Teachers need to be able to gather their own evidence and adjust their teaching practices for their students if the classroom community is going to equitably support all learners. Evidence-based teaching practices developed with that in mind account for context and student diversity while focusing on the whole student rather than narrowly defined learning as measured on standardized tests (Darling-Hammond et al., 2020; Nasir et al., 2020). The three evidence-based teaching practices we have found most integral to our approach are complex instruction with its focus on equity; productive struggle with its focus on building self-efficacy; and reflective teaching with its focus on leveraging classroom teacher expertise to improve teaching and learning.

Evidence for Using Complex Instruction to Promote Equity

In Chapter 3, we introduced complex instruction as an evidence-based teaching practice, emphasizing its focus on group-worthy tasks and status. Our focus in that chapter was on helping you understand what these teaching practices involve. Here we shift focus to describe evidence that complex instruction actually works and how it affects learners. Recall that status is socially determined and is influenced by race, class, and gender, but also by individual characteristics like reading ability, past math/science learning, and peer social influences. Status determines who believes they can enter academic conversations and contribute in group-learning contexts. Central insights of complex instruction are about how status drives who participates in on-task talk and to what degree, and what teachers can do to shift those dynamics toward equity for all learners. Studies of implementing complex instruction have consistently shown rates of on-task talk positively correlated with achievement gains, while studies focused on rates of on-task talk without complex instruction have not always shown such positive correlations, making it clear that the quality or type of talk is important (Bianchini, 1999; Boaler, 2006; Cohen & Lotan, 1997, 2014).

A great deal of research has shown that high-status students typically participate more in the classroom and during group work, ultimately learning and achieving more as a result (Cohen et al., 1999; Lotan, 2003; Perrenet & Terwel, 1997). Research evidence supports the use of two complex instruction teaching moves, "multiple-ability treatment" and "assigning competence," to shift status toward equity while engaging students in group-worthy tasks (Bianchini, 1997; Cohen & Lotan, 2014; Lotan, 2003; see Figure 4.1).

Highlighting contributions of lower-status group members benefits the entire group, including high-status members, as they compare mathematical and scientific ideas, consider alternate solution paths, and develop

Figure 4.1. Evidence-Based Complex Instruction Teaching Moves

- *Multiple-ability treatment* is consistently telling the class that many different skills are required for success on a group-worthy task and that no individual student has all these skills, but each student has some of them. This emphasizes that the participation of all students in a group is required for success.
- *Assigning competence* reinforces multiple-ability treatment and is done while a teacher moves through the classroom observing group interactions. The teacher looks for students who are marginalized in their group but have academic contributions to make. The teacher publicly calls out the student and contribution saying things like, "Tina's idea sounds like something the group should really think more about."

deeper understandings (Webb et al., 2002). There is good evidence that such teaching moves shift student thinking about who can and should contribute in mathematics and science discussions and lead to increased equity in the classroom and increased learning gains for all (Cohen & Lotan, 2014; Darling-Hammond et al., 2020). For example, students who received complex instruction shifted their beliefs, indicating that more of their peers were good at math and that there are many different ways to be good at it (Boaler, 2008). Comparable students who did not receive complex instruction did not make such shifts, and those broader perspectives of students who received complex instruction were related to larger achievement gains.

Published research evidence and our own experiences with teacher colleagues show that complex instruction, with its focus on equitable status, effectively shifts learner perceptions of who can and should participate in the classroom, and that equitable participation leads to increased achievement. We now turn to evidence for the importance of supporting students in productive struggle to build learner self-efficacy.

Evidence for the Importance of Productive Struggle

Educational researchers and organizations in mathematics and science teaching and learning have proposed practices that support students' learning in these disciplines (Ball & Forzani, 2011; Forzani, 2014; Grossman & McDonald, 2008; McDonald, Kazemi, & Kavanaugh, 2013; NCTM, 2014; National Governors Association and Council of Chief State School Officers [NGA & CCSSO], 2010; NGSS Lead States, 2013). For example, researchers in the "Tools for Ambitious Science Teaching" group (http://ambitiousscienceteaching.org/us/) and the National Council of Teachers of Mathematics (NCTM, 2014) articulated research-informed teaching practices that support ambitious science and mathematics instruction. These practices served as a framework for our approach. As we worked with our teacher colleagues as learners in relevant activities, we implemented these practices and then discussed them with teachers to support their planning and implementation of this instructional approach in their own classrooms. As we have continued to refine and study our approach, we have noticed that some teaching practices were particularly relevant to our experience. One practice that clearly stood out, and that connects the other practices, is supporting productive struggle in learning.

Cognitive learning theorists have highlighted the role that struggle plays in learning for years (e.g., Dewey, 1929; Piaget, 1960; Skemp, 1971; Steffe, 1991; von Glasersfeld, 1991). Productive struggle involves using prior knowledge to persevere through and make sense of problems or tasks. Practices that support productive struggle can be motivators that encourage students to move through moments when they are "stuck" in the process of working through a solution. Some learning theorists have found that even when efforts do not result in a successful solution, what they term *productive failure*, the act of working through the solution can have a positive impact on learning (Kapur, 2008, 2010, 2011; Kapur & Bielaczyc, 2012; Schwartz & Martin, 2004).

Although useful, supporting productive struggle in learning is not a simple practice. A variety of instructional features that support productive struggle in learning are suggested in the literature (Valentine & Bolyard, 2018; see Figure 4.2). Most of the literature discussing features that support the practice of engaging students in productive struggle is situated in mathematics instruction, likely because this is a specific practice listed in NCTM's (2014) recommendations of effective teaching practices.

While "productive struggle" does not appear as a core practice for ambitious science teaching, it is clear that it connects to those practices, particularly when looking at the characteristics of classrooms that engage in ambitious science teaching as described by the "Tools for Ambitious Science Teaching" group (see Figure 4.3). Such characteristics align with

Figure 4.2. Instructional Features that Support Productive Struggle

- Establishing a learning environment that values perseverance and normalizes failure (e.g., Kapur, 2010; Warshauer, 2015a, 2015b).
- Engaging students in high-level tasks (e.g., Stein et al., 2009).
- Valuing students' ideas and contributions (e.g., Stein et al., 2009; Warshaurer, 2015a).
- Scaffolding students' sense-making (e.g., Barlow et al., 2018; Engle, 2006; Franke et al., 2015; Townsend et al., 2018)
- Focusing on justification and explanation (e.g., NCTM, 2014, Stein et al., 2009)
- Encouraging and positioning students as contributors to the learning process (e.g., Doerr, 2006; Engle, 2006; Franke et al., 2015)

Figure 4.3. What Happens in Classrooms Where Ambitious Science Teaching Occurs

- Students are engaged in complex, challenging tasks while creating explanations of what they are discovering.
- Teachers and students are participants in a learning community that promotes sharing and questioning ideas as they move toward deeper understanding.
- Teachers encourage students to be active participants and press students for evidence-based explanations—scaffolding, as needed, to enable all to access learning.

instructional features highlighted as supporting productive struggle in learning, and the evidence base for their impact on learning is strong.

In our own work, the clear relevance of productive struggle prompted us to focus on more deeply understanding our teacher colleagues' perspectives related to this practice. Our teacher colleagues described experiencing discomfort, ambiguity, and moments of impasse during their professional development experiences with us. Importantly, they noted that the result was new learning and feelings of accomplishment. They used the language and ideas of productive struggle as they described these experiences. Particular features of our professional development design influenced their understanding of productive struggle: being able to experience design-based activities as learners and working through solutions with heterogeneous groups of their colleagues. As they became more comfortable with the approach, participants realized it was okay to not know or to take risks to try something new. Struggle became a normal part of the process of learning about content, teaching, and themselves. They also developed different perspectives on their students, as they were able to see the classroom from their students' viewpoint. They realized how students might engage in productive struggle and how they might support them in the process.

Evidence for the Value of Reflective Teaching Practices

In Chapter 3, we described reflective teaching practice as conceptualized by Dewey and Schön. Remember that this involves both reflection-in-action as teaching occurs and reflection-on-action when time allows careful consideration of multiple sources of information about how teaching practices impact learners. Dewey and Schön's perspectives have been taken up in teacher professional development (preservice and inservice) programs that focus on reflectivity, often in the context of action research. Some strengths of these approaches are their positioning teachers as active learners who develop local expertise in their classroom, their focus on multiple types of information useful for supporting reflection that leads to action, and their capacity to inform teaching practices that impact both disciplinary learning and equity.

The Association of Teacher Educators formed a Reflectivity Commission in 2004 "to provide a thorough and detailed investigation of the impact of teacher reflectivity," and the work of that commission was updated with more-recent empirical evidence (Manfra, 2019; Rigney et al., 2019). There is strong consensus and empirical support for the effectiveness and impact of reflective teaching practices. While most of the evidence is drawn from work with prospective (preservice) teacher candidates, there is also evidence for the impact of reflective teaching practices when implemented by current (inservice) teachers. The evidence spans a broad diversity of disciplines, including mathematics and science education. Reflective teaching practices improve teachers' subject-matter knowledge, pedagogical content knowledge, and integration of disciplinary inquiry into instruction, and positively shift teaching practices and teacher professional identities (e.g., Berry & Milroy, 2002; Doerr & Tinto, 2000; Geddis, 1996; Heaton & Mickelson, 2002; Jaworski, 2006; Katz & Stupel, 2016; Kyei-Blankson, 2014; Lampert, 1990, 2001). Reflectivity has been shown to help teachers uncover personal biases; understand student needs; and unpack social, political, and cultural contexts important to effective teaching and learning in their classrooms (Raygoza, 2016).

There are many examples of reflectivity impacting teaching and learning related to sociopolitical contexts and connecting with students' lives beyond the classroom (e.g., Addleman et al., 2014; Felton-Koestler, 2020; Jett & Cross, 2016; Ostorga & Estrada, 2009). In an example from mathematics, an algebra teacher engaged her students in investigating school food and health justice issues to connect mathematics content to sociopolitical issues (Raygoza, 2016). Her students showed new understandings of research, mathematics concepts, and social justice issues. She described how her teaching practices shifted toward more democratic interactions in the classroom. The Chicago River Project provides an example from science education where students from low-income neighborhoods focused on inequalities in their community while investigating illegal dumping of waste (Bouillion & Gomez, 2001). The project's findings support changing teaching practice to integrate hands-on activities and improving students' "ability to access information, form questions, share ideas, and analyze and compare data" (p. 888).

EVIDENCE BASE FOR CORE COMPONENTS OF OUR APPROACH

In addition to the evidence-based practices just discussed, our approach centers on (a) meaningful integration of science, mathematics, and literacy content; (b) the design process; and (c) mathematical modeling. Each of these areas has substantial empirical evidence supporting their effectiveness for enhancing teaching and learning.

Evidence for the Value of Content Integration

The strong evidence base for integrating mathematics, science, and literacy has influenced national and state education standards (e.g., Common Core State Standards for Mathematics, Next Generation Science Standards). These standards documents make it clear that mathematics is an essential tool for scientific inquiry, and science is a critical context for developing mathematical understandings (Doerr & English, 2006; National Research Council, 2006). Mutually reinforcing science and mathematics understandings while teaching either discipline is an important and effective interdisciplinary opportunity (Center for Educational Policy, 2007; NCTM, 2000). Integrating literacy practices to help students communicate and reflect on their learning further extends and grounds this opportunity for preparing students to meet tomorrow's challenges.

Before exploring the evidence base for content integration, it is important to understand what precisely we mean by integration. Many different forms of integration are described in education research, but forms with the strongest evidence base—those that we utilize in our work—are content-specific integration, methodological or process integration, and thematic integration (see Figure 4.4). All these forms of integration are present in project-based and problem-based learning approaches, which have strong evidence bases in educational research (e.g., Chen & Yang, 2019; Furtak et al., 2012; Lazonder & Harmsen, 2016). Recall that design-based learning is a form of problem-based and project-based learning that explicitly uses the design process and involves building something as (at least part of) a solution.

Evidence for the Value of Using the Design Process

The evidence base for the effectiveness of the design process in education comes predominately from science and engineering education (e.g., Fortus

---

Figure 4.4. Three Forms of Integration in Our Approach

- *Content-specific integration* is utilizing content standards from different domains; in our case those domains are mathematics, science, and literacy.
- *Methodological or process integration* is using real-world tasks and activities from those domains to engage in mathematical, scientific, and literacy practices.
- *Thematic integration* is contextualizing learning in a theme such as environmental stewardship, which allows relevant mathematics, science, literacy, and content from other disciplines to be explored in a meaningful context.

Figure 4.5. Characteristics of Good Designers

- Tolerate ambiguity and uncertainty resulting from design as inquiry
- Use systems thinking
- Make decisions
- Think and work in teams

et al., 2005). Dym et al. (2005) assert that the design process itself is a complex cognitive process and that developing design thinking skills enables students to develop the characteristics of good designers shown in Figure 4.5.

Research in this area draws on the evidence base for project-based learning, problem-based learning, and most directly on what is called Learning by Design (e.g., Barrows, 1985; Kolodner et al., 2003; Prince, 2004; Thomas, 2000). Learning by Design is grounded in the evidence base for how people learn (National Academies of Sciences, Engineering, and Medicine, 2018) and focuses on deep learning of science content along with complex cognitive, social, and communication skills and the cultural contexts in which they develop.

One of many potential examples of this evidence base comes from researchers at the Learning Research and Development Center at the University of Pittsburgh (Mehalik et al., 2008). These authors compared middle grade student learning of the same science content when taught through Learning by Design or through a more traditional "scripted," or teacher-directed, problem-based learning approach. Learning by Design students achieved two times larger learning gains compared to their peers receiving teacher-directed instruction, and those gains were largest for lower-achieving African American students. While these authors point out that national science education standards call for the use of mathematics in all aspects of scientific inquiry, we have found that too often, problem- and design-based learning projects have very little meaningful mathematics content.

Evidence for the Value of Mathematical Modeling

The evidence base for the effectiveness of mathematical modeling comes predominantly from mathematics education (e.g., Sokolowshi, 2015). Scientists, engineers, and other professionals often do not use mathematics in ways that mirror how it is taught in school. Rather than follow outlined steps and procedures, they use it more flexibly: quantifying variables, making judgments about what is important, and creating and revising models for messy problems. Experiencing mathematical modeling in school can allow students to use mathematics in more flexible, creative, and powerful ways that are more in line with how it is used in professions (Zawojewski, 2016).

In fact, research has found that the skills and understandings students draw upon and develop in mathematical modeling contexts are those that are used in fields such as engineering and business management (Lesh & Yoon, 2007). These include (a) making sense of complex situations; (b) working collaboratively with heterogeneous groups; (c) being able to design, implement, evaluate, and revise solutions; and (d) being flexible in thinking.

Several studies have examined students' cognitive processes while engaged in modeling activities (e.g., Blum, 2011; Galbraith & Stillman, 2006; Stillman & Brown, 2021). The complexity of these skills (e.g., connecting mathematics to the real world, using problem-solving strategies) means that modeling tasks are very cognitively demanding. To support students' ability to transfer their understanding and skill to modeling tasks, they must work with a range of contexts and content domains. Despite the challenges of modeling tasks, students can be successful in an environment that supports their learning and independence. Teacher scaffolding that emphasizes learner independence while helping them move through an impasse is key to students' success in modeling activities (Kaiser, 2017). Further, metacognitive activity can help students to "recognize their problems and errors on their own while monitoring" their actions as they construct and reflect on their models (p. 287).

The strength of this evidence base prompted Common Core State Standard for Mathematical Practice Number 4, Model with Mathematics, a practice that requires the application of all other Standards for Mathematical Practice (Common Core State Standards Writing Team, 2013). The evidence base for the power of mathematical modeling also supports Next Generation Science Standards that focus on students engaging in iterative cycles to develop and refine models to compare predictions with the real world and gain insights into what they are modeling.

The National Council of Teachers of Mathematics published a book on mathematical modeling in 2016 that describes *mathematical modeling* (using mathematics to represent the real world) and distinguishes it from *modeling mathematics* (using models to represent mathematics; NCTM, 2016). That book described how mathematical modeling promotes student thinking and student-generated algorithms, rather than students simply remembering teacher-generated algorithms. The authors wrote that "the mathematical understanding that students develop when they are 'thinking' and not merely 'remembering' is more powerful, more flexible, and longer lasting" (Gann et al., 2016, p. 106).

EVIDENCE BASE FOR BRINGING IT ALL TOGETHER

So far in this chapter, we have taken an analytic approach to describe the various components of our model, focusing on educational research that supports their effectiveness. We began with evidence-based teaching practices such as complex instruction, productive struggle, and reflective teaching. Then we addressed the evidence base for each of the core components of our approach: content integration, the design process, and mathematical modeling. We conclude this chapter by discussing evidence that we have developed while applying our model in professional development with teacher colleagues over the years.

Since 2011, we have produced six research publications and seven academic conference papers describing our approach and the evidence for its effectiveness (Bolyard et al., 2018; Cairns et al., 2018; Curtis, Bolyard, Cairns, & Loomis, 2021; Curtis, Bolyard, Cairns, Loomis, Mathew, & Watts, 2017a, 2017b; 2017c; Curtis, Bolyard, Cairns, Mathew, Loomis, & Watts, 2017; Curtis, Cairns, & Bolyard, 2020; Curtis, Cairns, Bolyard, Loomis, Watts, Mathew, & Carte, 2016; Curtis, Cairns, Bolyard, & Walker, 2018; Curtis & Georgieva, 2011; Curtis, Georgieva, Cairns, & Solley, 2013; Georgieva et al., 2013). Most of our work has focused on the impact of our approach on our teacher colleagues' content knowledge, attitudes, and self-efficacy, while our teacher colleagues have described the impact it has had on their students. We describe two examples of our contributions to this evidence base here.

In 2013, we published a paper that we had presented at the American Society for Engineering Education based on our earliest work in this area (Georgieva et al., 2013). At that point, we had not yet integrated complex instruction or come to understand the critical importance of productive struggle, but we now recognize that the foundations for those effective teaching practices were there even then. In that early piece, 23 of our teacher colleagues engaged in cutting-edge engineering research at West Virginia University, attended an International Summer Energy School at the University of Birmingham (United Kingdom), and developed design- and problem-based learning units based on those experiences to deliver to their students. This work was with high school teachers, and we used validated content knowledge tests, including the Force Concept Inventory (Hestenes et al., 1992) and the Calculus Readiness Assessment (Pyzdrowski et al., 2013). We also created a chemistry test based on the AP Chemistry exam and assessed teachers' attitudes and understanding of design- and problem-based learning, mathematics-science content integration, engineering, energy renewal, and environmental sustainability at multiple time points. We interviewed all participating teachers several times throughout the program and analyzed the lesson plans they created.

We found that engineering research and international experience around sustainable forms of energy, along with participating in engineering design activities, were transformative for our teacher colleagues. Participating teachers came to better understand engineering and the design process, and made connections among mathematics content, science content, and real-world applications to socially relevant problems that they previously did not recognize. This transformation carried into their classrooms through design- and problem-based learning units, and through increased advocacy for sustainable energy solutions and STEM education and career paths.

More recently, in 2021, we presented at the American Educational Research Association conference a four-part model for focusing teacher professional development and evidence for its effectiveness (Curtis et al., 2021). We described how we used iterative design and redesign to address "the engineering problem" of building teacher pedagogical content knowledge for mathematics and science. The importance of productive struggle and mathematical modeling figured prominently in this paper, and complex instruction was now an important part of our work with teacher colleagues. Our model for professional development included the following steps:

1. Identify mathematics and science learning opportunities utilizing established standardized assessments (we used the Diagnostic Mathematics and Science Assessments for Middle School Teachers tests [DTAMS; Saderholm et al., 2010] and the Teacher Efficacy and Attitudes Toward STEM surveys [T-STEM; Friday Institute for Educational Innovation, 2012]).
2. Engage teachers in productive struggle as learners in integrative design-based tasks requiring that knowledge and including mathematical modeling for prediction prior to building and testing designs.
3. Require teachers to design, implement, and redesign design-based lessons addressing related learning opportunities with their students.
4. Evaluate teacher learning through observations, interviews, and pre-post testing with standardized assessments, and redesign tasks to support further learning.

We described data from 25 teacher colleagues who experienced our model and 22 comparison teachers who did not. Participating teacher colleagues demonstrated increases in their confidence to teach the design process, use of technology with students, and knowledge of STEM careers more than teachers from similar schools. Participants valued active learning, collaborating with peers and experts around teaching, focusing on content across subject areas, and focusing on student learning.

A compelling issue across annual interview data was productive struggle—in particular, that of teachers—seen in their comments about themselves, student effects, and responses from parents and administrators. Our teacher colleagues experienced productive struggle authentically, and their learning opportunities were like those of their students. Design-based learning provided an experimental framework that became familiar to them and enabled further, richer experimentation that was targeted at understanding learning opportunities and could be adapted for use in their classrooms. We found sustained engagement with our teachers critical, and teachers were more open to addressing their own learning opportunities in the context of design projects focused on similar learning opportunities for their students, learning opportunities that happened to overlap with content knowledge our teacher colleagues needed to develop more deeply themselves. An expanded version of this paper is currently under review for publication.

We have now described the existing evidence base, from our own work as well as the larger educational researcher community, for the components of our model separately and for the use of the full model in our provision of teacher professional development. We hope we have accomplished our goal for this chapter, which was to help you understand why we believe this approach to be so powerful for facilitating deep and integrated teacher and student learning. In the next chapter and beyond, we turn our focus to helping you understand how to put this approach into practice.

CHAPTER 5

Personifying Best Practices

You are reading this book because you care. You care about your students and you care about being a great teacher. We know from education research that teacher quality impacts student learning, and that student achievement rises dramatically when students have good teachers who care. Chapter 4 described the evidence supporting the effectiveness of the approach we share in this book as a set of best practices. But educational researchers routinely point out the difficulty of linking educational theory, such as our approach, to educational practice, or what teachers do in their classrooms.

How do you take the theory we describe and put it into practice effectively with your students? Our teacher colleagues have found it valuable to switch back and forth between considering learning experiences using this approach as learners and as teachers—for example, putting on the "teacher hat" to consider what content standards can usefully be integrated, putting on the "learner hat" to consider which parts of learning experiences focused on that content are likely to be challenging for students, and then again putting on the teacher hat to think about what instructional supports can best help students learn that content.

This chapter is focused on helping you link theory to practice and personify best practices. We begin with a real-world example from one of our teacher colleagues and then focus on describing how her teaching personified best practices. As you read this, we encourage you to reflect on what you can learn from focusing on the perspectives of both students and teachers.

(RE)DESIGNING INDUSTRIAL FARMING IN YOUR STATE

Ms. Williams joined our professional development in her first year teaching 7th- and 8th-grade mathematics and science in a school located on the edge of a small city. She decided to put our approach into practice while she was in the middle of a traditional unit on ratios and proportions. She wanted to shift her practice and connect to her students' everyday lives. Knowing that many of her students came from rural areas, she decided to draw on their background knowledge of farming practices and find ways to link that context to meaningful mathematics, science, and literacy through design-based learning activities.

She began by asking her 7th- and 8th-grade students what they knew about livestock farming. She asked her students to be on the lookout for ways to use mathematics as they explored the science and ethics of industrialized farming practices. She emphasized that no single student would have all the answers or skills needed to explore this, but that everyone in their group had important things to contribute. Each member of the group had an assigned role (e.g., researcher, recorder, or organizer) that changed from day to day, and all students were responsible for keeping a notebook to record what they did and learned. As Ms. Williams watched students work in groups, she occasionally highlighted something one or another student said, especially when a student who was less vocal in the classroom contributed.

Ms. Williams's students learned that a problem with industrialized farming is the number of animals housed in the space provided. When there is not enough space, conditions can be inhumane, and because the waste produced is concentrated in a small space, it becomes an environmental problem affecting the air, water, and land that humans use as well. When there is enough space, conditions can be more humane, and environmental impacts can be mitigated. Students were intrigued. What her students did not know was that Ms. Williams had identified several content standards and associated learning goals that students would meet as they explored this topic in her classes (see Figure 5.1).

As the importance of space in the "problem" of industrialized farming became clearer, students focused on determining a solution related to the land needs of the state they live in. Student groups selected one domestic animal to be the focus of their study and researched what they needed to know to calculate proportions and unit rates for "proper" facilities to house those animals. They researched the amount of meat products consumed in their state and developed calculations and graphic displays, revising their solutions as they compared calculations with what they learned through Internet research. Finally, student groups prepared posters, made presentations to the class, and wrote reflections on other groups' presentations about the state's land needs for humane industrial farming to satisfy the population's meat consumption.

As students worked, Ms. Williams paid close attention to group functioning and the mathematics and science content her students struggled with, making notes to herself. She analyzed this information and used what she learned to guide short lessons and other supports to focus the class on specific content as needed across the 5 days devoted to this unit. Ms. Williams scored student presentations at the end of the unit using a rubric she developed that was tied to student learning goals based on

Figure 5.1. Standards and Learning Goals in Industrial Farming Design-Based Lesson

---

CCSS-M Mathematics Content and Practice Standards

- 7.RP.2: Recognize and represent proportional relationships between quantities.
- 8.EE.B.5: Graph proportional relationships, interpreting the unit rate as the slope of the graph, and compare two different proportional relationships represented in different ways (e.g., compare a distance-time graph to a distance-time equation to determine which of two moving objects has greater speed).
- 8.EE.C.7: Solve linear equations with one variable.

NGSS Science Standards

- MS-ESS1-4: Apply scientific principles to design a method for monitoring and minimizing a human impact on the environment.

CCSS-ELA Standards

- SL.7.4: Present claims and findings, emphasizing salient points in a focused, coherent manner with pertinent descriptions, facts, details, and examples; use appropriate eye contact, adequate volume, and clear pronunciation.

7th- and 8th- Grade Learning Goals:

- I can recognize and represent proportional relationships between quantities.
- I can decide whether two quantities are in a proportional relationship.
- I can identify the unit rate (constant of proportionality) in tables, graphs, equations, diagrams, and verbal descriptions.
- I can represent proportional relationships by equations.
- I can explain what a point (x,y) on the graph of a proportional relationship means in terms of the situation.
- I can explain what (0,0) and (1,r) mean where "r" is the unit rate.

8th-Grade-Only Learning Goals:

- I can write a linear equation based on real-world data.
- I can solve linear equations using different values.

---

the content standards she initially selected. As she reviewed student reflections, posters, and rubric scores from their presentations, Ms. Williams thought through what seemed to work well and how she might improve parts of these lessons in the future.

CONNECTING THEORY TO PRACTICE

In Chapter 3, we presented the full theoretical model for our approach. As you know, the key components of this model include (a) using the design process with mathematical modeling; (b) integrating mathematics, science, and literacy content standards; and (c) engaging in evidence-based teaching practices (i.e., complex instruction and reflective teaching) to support productive struggle and equity in the classroom. Consider how Ms. Williams's industrial farming unit incorporated each of these components.

Design With Mathematical Modeling

Without explicitly labeling the stages of the design process, Ms. Williams first engaged students in the Empathize stage as she asked them to consider industrial farming from the perspectives of farmers, other people impacted by farming in the area, consumers who use the meat farming provides, and even the animals. This approach presented multiple entry points for students to engage with the content, an important feature of group-worthy tasks.

Ms. Williams skillfully engaged students in critical thinking about what they were learning, and they shifted to the Define stage as they identified problems related to the amount of space available, humane animal treatment, and environmental impact. At this stage, students developed a problem statement (e.g., How can we design a farm structure that can house at least x number of sheep with adequate space per sheep at a reasonable cost?). Ms. Williams's asking students to look for ways to use mathematics made it natural for her to bring mathematical modeling into the Ideate and Prototype stages as they developed, compared, and revised solutions leading to understanding how much space was necessary for humane treatment of animals and appropriate environmental stewardship while addressing the consumption needs in the Test phase. The Communicate phase involved creating and presenting posters, as well as reflecting on what students had learned from their peers' presentations.

Integrated Content Standards

Ms. Williams integrated content standards for mathematics, science, and literacy, naming those standards and including them in her lesson plans. Notice how the standards fit together naturally within the context of the design-based learning unit on industrial farming. The mathematical tools defined by the mathematics standards were needed to address the science standard, and the science standard provided a meaningful context within which to use those mathematical tools in meaningful ways. It was natural to use spatial measurements, count the number of animals in a space, mathematically model and graph relationships between numbers of animals and space, and use those mathematical models to consider how to monitor and minimize environmental impact while considering consumer needs.

The requirement for students to create posters and present their findings connected to a literacy standard. Ms. Williams also asked students to write reflections. The content standards Ms. Williams focused on led directly to learning goals that Ms. Williams connected explicitly to authentic assessment of student work through a rubric she created. If she were able to collaborate with an English language arts teacher in future deliveries of this unit, students could make meaningful connections across classes.

Complex Instruction

The industrial farming group exploration qualifies as a group-worthy task as described in Chapter 3. It was open-ended and included multiple ways that students could enter or contribute to the task. It was integrated with meaningful content through Ms. Williams's focal content standards and student learning goals. It was enacted through collaborative group work where both the group as a whole and students individually were held accountable for learning. Ms. Williams included clear criteria for what groups and individual students needed to create and record in their notebooks, posters, presentations, and reflections. Ms. Williams also used two teacher moves from complex instruction described in Chapter 4 for effectively supporting equitable learning: multiple-ability treatment and assigning competence.

Multiple-ability treatment in this example was what Ms. Williams was doing when she emphasized that no single student would have all the answers or skills they needed for this task, but that everyone in the group had important things to contribute. Keep in mind that this was a way of thinking that Ms. Williams consistently reinforced rather than just a single statement at the start of the unit. The rotating roles Ms. Williams assigned students emphasized this way of thinking as well. Ms. Williams assigned competence through highlighting something a student said that made an important contribution to their group or the whole class. She could also have done this by sharing an entry in a student notebook to assign competence to a student whom she might not have observed making contributions during group work. As Ms. Williams recognized students struggling with some of the content, she worked to scaffold that struggle to be productive through mini-lessons and other supports she thought would help her students. We will discuss scaffolding such as this in more detail in Chapter 8.

Reflective Teaching

Ms. Williams engaged in both reflection-in-action and reflection-on-action as described in Chapter 3. She also used two types of authentic assessment: formative and summative. *Authentic assessments* are different from traditional knowledge quizzes or tests because they arise naturally from learning activities, such as when Ms. Williams observed students working in groups or making their presentations. Authentic assessment is grounded

Figure 5.2. Sample Presentation Rubric Item Assessing Learning Goal 1

Learning Goal 1: I Can Recognize and Represent Proportional Relationships Between Quantities

Misconceptions	Beginning	Proficient	Sophisticated
Representing relationships that are not proportional in a proportional format.	Have written at least half correct proportional relationships and can apply them to industrialized farming situation.	Have written correct proportional relationships, can apply them to industrialized farming situation, and can mostly explain the relationships in detail to the class and how they were used.	Have written correct proportional relationships, can apply them to industrialized farming situation, and can fully explain them in detail to the class and how they used them.

in the doing of a task and can be facilitated by structured observation or scoring rubrics, such as when Ms. Williams devised a rubric to score student presentations (see Figure 5.2). *Formative assessment* is used to guide instruction as it unfolds (e.g., observing group work to determine topics for mini-lessons on content), while *summative assessment* is used to understand how well instruction has fostered student learning (e.g., using a rubric to score student presentations).

Reflection-in-action should be grounded in authentic formative assessment and help you make instructional decisions in the moment. This is what Ms. Williams was doing as she paid attention to student discussions around content. Her reflection-in-action helped her determine the content for which students needed additional support and scaffolding. When she noticed and highlighted a student contribution or when she noticed students struggling and supported their learning in the moment, she used authentic formative assessment to guide reflection-in-action.

Her reflection-on-action came later as she considered how well her students were progressing toward the learning goals she had set for them. She did this daily as she considered how to guide her teaching practices for the next day. She also did this at the end of the unit as she reflected on the entire experience and considered changes she would make in the future. Ms. Williams's reflection-on-action drew both from her formative assessments while teaching and from summative assessment of students' notebooks, posters, presentations, and reflections on what they learned from other groups. When Ms. Williams developed mini-lessons or other learning supports, she was using reflection-on-action and formative assessment. After the unit was done, when Ms. Williams considered how she might change these lessons next time, she was using summative assessment and reflection-on-action.

Ms. Williams also used the design process to think about her instruction as a "solution" to the "problem" of fostering equitable student learning related to these content standards. As she continues to revise and improve her lessons, she will be using the design process to improve student learning over time. The earlier stages of the design process are a particularly useful moment to put on your student hat to consider what your students need to learn and what might be particularly challenging areas for students. It is important to Ideate multiple potential solutions (e.g., learning activities or parts of learning activities) and to Prototype and Test those solutions in the classroom. As you do that, reflection-in-action and reflection-on-action help you consider how and what to continue using or to revise next time around. Communicating with your teacher colleagues and network what you are learning, what is working, and what your challenges are further helps you consider what and how to refine your practice.

As we conclude this chapter, we want to point out another valuable focus to consider for reflection-on-action through the design process: equity, both in the classroom and beyond. Group-worthy tasks and the teacher move from complex instruction focus on equity in the classroom, and this is where Ms. Williams's focus was in this lesson. However, as Ms. Williams continues to develop as a teacher, we are hopeful that she will also facilitate her students' focus on equity beyond the classroom toward larger societal issues that impact their communities.

As mentioned in Chapter 4, reflective teaching practices have sometimes been used to unpack social, political, and cultural contexts to bring a critical equity focus into students' lives. You may be thinking of Gutiérrez's equity framework introduced in Chapter 3. Her critical axis with power and identity equity dimensions is particularly relevant here. Students should see themselves and understand others through the curriculum, and they should have voice and agency in the classroom and beyond. A unit such as this one on industrial farming could easily lead into how farming practices affect lower-income, urban versus rural, or marginalized groups differently. Such a focus could frame discussions within this unit or be expanded in larger additional curricular units across months, semesters, or even the entire year. Students could learn relevant mathematics, science, and literacy content in the context of researching real-world societally relevant aspects of food production, marketing, and corporate and consumer practices that directly impact their own lives. Students could even develop solutions and advocate for their use in their communities. Such interrelated and societally relevant content could fire the imaginations and social activism of students while at the same time solidifying deep, meaningful learning.

Part II

MAKING IT REAL

CHAPTER 6

Design and Mathematical Modeling— From Artifacts to Processes

Dan Meyer (2009) posted to his popular blog on math education a problem that he described as his "Da Vinci code." The post showed a figure similar to Figure 6.1, which depicts two different grocery store checkout lanes. The white boxes represent customers, and the number inside each indicates the number of items that customer has in their shopping cart. Meyer asks, "All other things being equal, which lane is the fastest?" You and your students have probably asked the same thing more than once while deciding which checkout line to get in.

This is a wonderful example of a problem that requires an understanding of the underlying process. In fact, this can also be used as an example of designing a process to make such a decision for a shopper at any store with checkout lines. However, developing the rules to decide which line to join requires an understanding of how the checkout process works. Dan did research himself by watching people shop over a 90-minute period and taking notes of how long it took for someone to get through the checkout line as a function of the number of items per customer. He even got the store manager to provide some data from the store. One way to analyze these data is to plot the time taken to get through the checkout line as a function of the number of items per customer, as shown in Figure 6.2.

This graph can be used to create a simple mathematical model that can help not only to answer Dan's initial question but also to design a decision-making process. A careful consideration of the prediction line in Figure 6.2 suggests that on average every item takes about 3 seconds and there are roughly an additional 40 seconds required for the rest of the process, including paying, finishing bagging, and so forth. This model can then be used to estimate how long the lines would take. Cash Register 1 has one customer with 19 items in their cart, and the estimated time would be 19 times 3 seconds per item plus 40 seconds for one customer, which is 97 seconds. Cash Register 2 has four customers with a total of 15 items, and therefore the estimated time would be four customers times 40 seconds each plus three seconds times 15 total items, which is 205 seconds.

Figure 6.1. Grocery Store Line

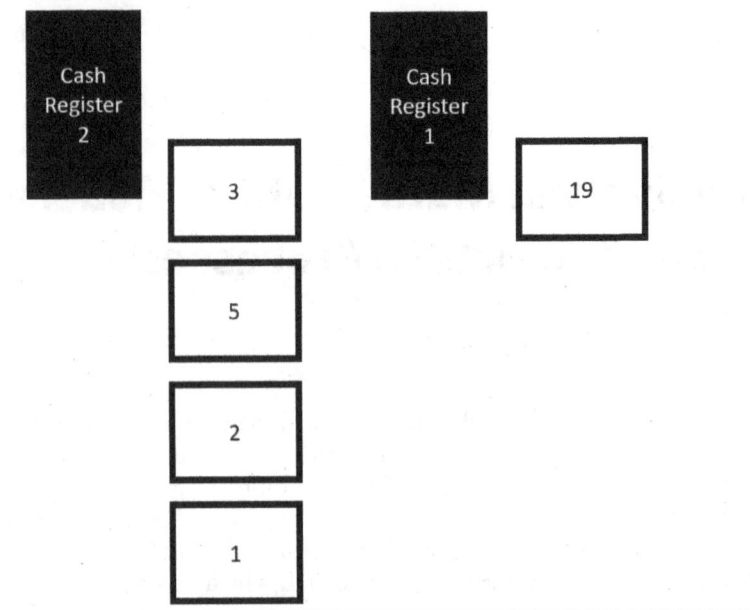

Source: Adapted from Meyer (2009).

Figure 6.2. Graph of Time at Checkout Versus Number of Items per Customer

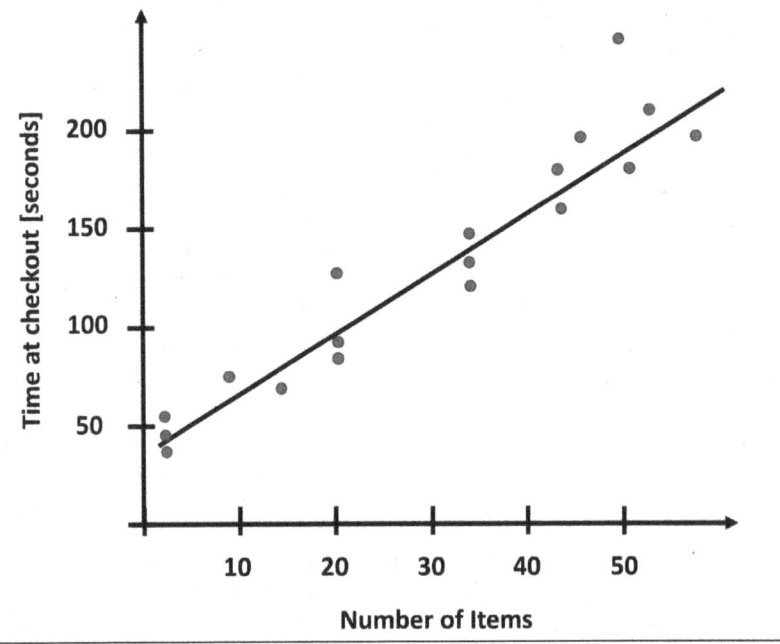

The process of choosing could therefore be: count the number of customers in each line, count the number of items each customer has, perform the calculation for each line, and choose the line with the shortest estimated time. The activity does not have to end there. There are many things that could be taken into account. For example, what sources of variability or error in prediction might emerge? What about customers who are getting cash back from their credit card or paying by check? How about self-scanning checkouts? You could also run simulations in your classroom with some students working at a cash register and others posing as customers. Design thinking to develop processes such as in this grocery story example is related to different affordances than are present in design activities focused primarily on artifacts. In this chapter, we discuss these affordances, how almost all design activities have at least some level of process thinking embodied in them, and how this focus on process can provide a gateway to developing systems-level learning.

AFFORDANCES FOR SYSTEMS-LEVEL LEARNING ACROSS THE ARTIFACT-TO-PROCESS CONTINUUM

We return here to the idea of affordances introduced in Chapter 2, where we defined affordances as teaching challenges that our approach helps to address. Chapter 3 focused on two key affordances: fostering equity in the classroom and supporting productive struggle. Here, we focus on how our approach provides affordances to support systems-level learning through design activities that vary in how focused those activities are on designing artifacts as compared to designing processes. Figure 6.3 shows a continuum of activities from those that focus exclusively on designing artifacts, such as a kit for building molecules, to activities that focus exclusively on designing processes, such as choosing the best checkout line in the grocery store. We focus the remainder of this chapter on understanding this continuum and how the design activity examples in Figure 6.3 have different affordances for facilitating systems-level learning.

Fostering systems-level learning is arguably a peak goal for education. The National Research Council (2012) wrote that cross-cutting concepts "help provide students with an organizational framework for connecting knowledge from the various disciplines into a coherent and scientifically based view of the world." It included "systems and system models" as a cross-cutting concept focused on "defining the system under study, specifying its boundaries and making explicit a model of that system" (pp. 83–84). The ability to engage in systems-level thinking is increasingly recognized as a key value in the workplace and an important goal for education.

A key affordance of design-based learning activities is that they are especially helpful in supporting systems-level learning. *Systems-level learning*

Figure 6.3. Artifact-to-Process Continuum of Design Activities

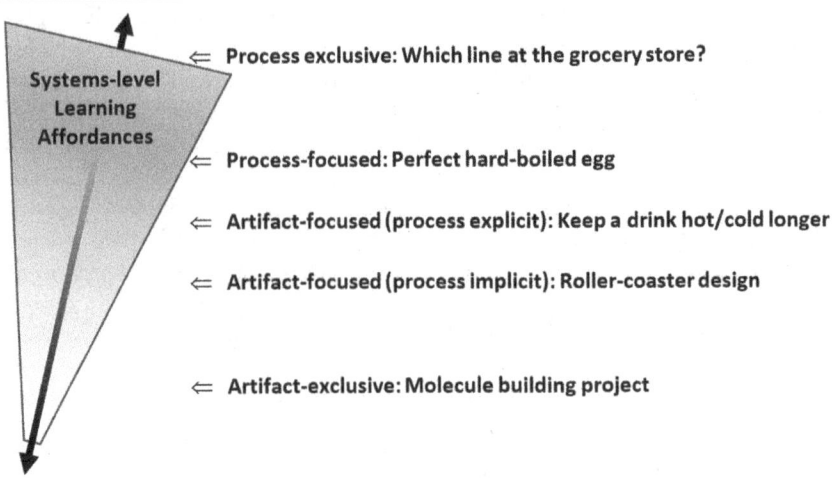

⇐ Process exclusive: Which line at the grocery store?

Systems-level Learning Affordances

⇐ Process-focused: Perfect hard-boiled egg

⇐ Artifact-focused (process explicit): Keep a drink hot/cold longer

⇐ Artifact-focused (process implicit): Roller-coaster design

⇐ Artifact-exclusive: Molecule building project

involves coming to understand how the world works in terms of relationships, patterns, and contexts that impact how groups of things and actions interact to give rise to the properties of complex systems (e.g., Capra & Luisi, 2014; Meadows, 2008). This should call to mind the distinction between analytic and synthetic thinking from Chapter 2. Analysis is particularly useful to simplify learning when initially approaching a complex system, but synthetic thinking is required to understand systems in the real world.

Systems are made up of both things and actions, which fits well with our focus in this chapter on designing artifacts (things) and designing processes (series of actions or events). Design activities that focus learners on making and redesigning artifacts provide different, but complementary, opportunities for learning compared to design activities that focus learning on foundational processes. Both are needed to foster deep systems-level learning, and the following sections focus on affordances for learning from design activities across the spectrum, from focusing exclusively on designing artifacts to focusing on designing processes.

Artifact-Exclusive Design Activities

We mentioned in the Introduction that there are far too many examples of work labeled as STEM or STEAM that do not make it clear which of the disciplines are being integrated and how or even if they are integrated in meaningful ways at all. An initially uncomfortable, but ultimately productive, experience with a group of our teacher colleagues is a good example of what we mean when we say an activity does not meaningfully integrate content across disciplines, or at the very least does not make cross-content-area connections explicit in a way that would be valuable for learners. Ms. Vaughan brought a "STEAM activity" that she had used in her science class as part of a unit on the structure of molecules. We described in Chapter 2 how she redesigned this unit to have more rich content integration through design-based learning.

As initially conceived, students used cocktail sticks and candy to make 3D models. Students built and shared their models with one another, but Ms. Vaughan was disappointed that students did not seem to learn as deeply as she had hoped. After Ms. Vaughan shared her activity in one of our professional development sessions, one of us asked a challenge question meant to focus the group on content connections that the activity did not capitalize on: "What makes this STEAM? In what ways is this more than a craft project?" Several of our teacher colleagues took exception to the term "craft project," and a lively discussion ensued.

Two categories of understandings emerged from that discussion: (a) the value (affordances) of bringing what we now call artifact-exclusive design activities into science and mathematics classrooms, and (b) the value of designing learning activities to capitalize on rich science, mathematics, and literacy content connections in addition to the building and making inherent in artifact-exclusive design activities. We agree wholeheartedly with our teacher colleagues that there is value in having students create, build, and make in mathematics and science classrooms, as you no doubt expect given how we have described design-based learning in this book.

The *affordances of artifact-exclusive design activities* include strengthening fine motor control, developing manufacturing and making skills with the range of materials used, problem solving through challenges that emerge in the building process, and considering the aesthetics of designs. When implemented in a group-learning format, artifact-exclusive design activities are a good way to build interpersonal skills and equitable learning norms for complex instruction. If the design process is introduced for design and redesign, artifact-exclusive design activities can also be fun ways to begin understanding the design process itself.

All of these are foundational things you need learners to develop as you begin implementing our approach. It is not that we do not see value in artifact-exclusive design activities. Rather, we see them as having a somewhat limited place in mathematics and science classrooms because more cross-content integrative activities are more efficient for facilitating deep systems-level learning about both artifacts and the processes that affect them in the real world.

Artifact-Focused (Process Implicit) Design Activities

Design activities focused on creating artifacts include all the affordances for learning that artifact-exclusive design activities include and more. The roller-coaster design activity is a good example. This activity is focused on building a thing, a paper roller-coaster track that keeps a marble going for a certain amount of time.

As students work to design, build, and redesign the roller-coaster track for the marble to follow, they learn all the things that come with building and making, as described in the previous section. As they try out different configurations to see what effects different track components (straight or curved ramps sloped up or down, loops, funnels, etc.) have on the marble, they learn how the physical properties of the track interact with the marble to influence its speed. They struggle productively to find configurations of track that will keep the marble rolling for the required amount of time. Students develop understandings of the roller coaster as a system involving different track components and a marble, and they focus on how redesigning various parts of the system influence the system's overall behavior in terms of time taken for the marble to traverse the track.

Teachers can use complex instruction to keep everyone contributing equitably as they naturally ask themselves questions and consider how to find the answers. A typical series of questions a group of students might ask themselves includes: Which types of track make the marble go too fast or even fly off the track? Which types of track slow the marble down, but not so much that it stops completely? How do different series or orders of track components affect the marble differently? Why does timing the marble on a small section or single component of track by itself not make the marble take the same amount of time as when that section is part of a longer series of track components? What matters, and how can the next design be better? These kinds of questions can be explored by students through trial and error, while they struggle productively and develop implicit understandings of the underlying processes grounded in real-world experience.

The underlying processes, physical mechanics involving gravity, velocity, mass, and momentum, are not an explicit part of this kind of design activity. However, these *artifact-focused (process implicit) design activities* provide important opportunities for teachers to scaffold student learning toward focusing more explicitly on underlying processes connected to content standards. We discuss scaffolding student learning in more detail in Chapter 8, and an important part of reflective teaching is adjusting your scaffolding to keep learners engaged and progressing toward understanding the content connections you need them to make. Often, those content connections involve understanding foundational processes and how those processes impact the system as a whole.

Solutions to design challenges such as the roller coaster are much more effective when students learn about the underlying processes, and that means an important affordance of this kind of activity is that it motivates students to be interested in learning about those foundational processes. The same is true of other artifact-focused design activities where the processes that impact the design are not explicit, such as the gingerbread house and skater ramp activities described in Chapter 9.

Artifact-Focused (Process Explicit) Design Activities

Artifact-focused design activities can also be crafted by reflective teachers to explicitly focus students on foundational processes. When you do this, you enlarge students' systems-level thinking to incorporate foundational processes as part of their understanding of the system they are designing. A good example of an artifact-focused design activity where the process is an explicit focus is the design challenge asking students to help teachers at their school figure out which cup really keeps iced drinks cooler longer, which we describe further in Chapter 7.

The reason the process is explicit in the keep-a-drink-cold example is that you cannot consider how to keep a drink cold (or hot) without considering the underlying process and physical mechanics of heat transfer through the involved materials. It is, in a real sense, the process of heat transfer that students are designing an artifact to control. That is why an important affordance for learning through *artifact-focused (process explicit) design activities* such as this is the ability to focus learners on foundational processes. It is worth noting that focusing on processes through design-based learning is uncommon in K–12 education even though it has great potential to improve systems-level thinking and learning.

As discussed further in Chapter 11, a series of linked design activities can be crafted to first get students interested in foundational processes (motivation for learning affordance) and then focus more explicitly on those processes. When you do that, you are capitalizing on the affordances of these kinds of design activities to facilitate learning all the same kinds of things that artifact-exclusive and artifact-focused (process implicit) design activities support, while additionally capitalizing on the affordance to focus on foundational processes. This also facilitates more sophisticated systems-level thinking as students incorporate foundational processes into their understandings of the systems they are designing.

Process-Focused Design Activities

Some design activities involve artifacts but their primary focus is on a process rather than the artifact itself. A good example of this is the perfect hard-boiled egg activity described in Chapter 1. This design challenge is about creating a process that affects an artifact. What is the best process through which to make the perfect hard-boiled egg?

Process-focused design activities have the important learning affordance of focusing students on the foundational processes that affect artifacts. In the hard-boiled egg example, these include the process of heat transfer and chemical processes related to how heat affects different materials such as the egg white and the egg yolk. Understanding foundational processes such as these is critical for systems-level learning and is at the heart of many important content standards.

The trade-off is that these design activities, and even more so process-exclusive design activities, do not have as many affordances for learning skills related to building and making as compared to artifact-exclusive and artifact-focused design activities. For this reason, it is important to use a diversity of design activities across the artifact-to-process continuum shown in Figure 6.3 so that learners develop the full range of understanding and skills that design-based learning affords.

Process-Exclusive Design Activities

Most often, engineering and design in K–12 education is focused on making and building things. A rarely understood affordance of the design process is that it can be applied to processes independent of artifacts. Dan Meyer's example about deciding which line to get in among several options to check out at the grocery store is a good example (Meyer, 2009). This example was so popular when he posted it on his blog that he was even invited to talk about it on the television program *Good Morning America*.

A transformative affordance of *process-exclusive design activities* is that they help learners see the applicability of design to decision-making in their daily lives. You could easily facilitate a discussion in your classroom about what challenges your students face in their lives. Many of the things likely to come out of that discussion could lead to design activities something along the lines of "how should I do something" or "what is the best choice in a particular situation": How should I study to have as much free time as possible and still do well on an upcoming test? How can I save money to help my family with expenses or to buy a gaming console? How should I decide which extracurricular activities to participate in at or after school? In fact, we hope you will apply this to your own decision-making and reflective teaching: How should I design tomorrow's instruction?

RETURNING TO THE GROCERY STORE

Dan Meyer (2009) explained why he loves this problem in the following terms: "It's my love for math, for mathematical reasoning, for the relentless deconstruction of something that seems simply intuitive into data, models, and computation." This deconstruction is key to systems thinking. Being able to think about the details and the whole simultaneously and using that flexibility of thinking is a powerful way to understand difficult problems and come up with creative solutions. After working on the grocery store activity to understand the process, students could then be ready for some challenging design activities that allow them to think creatively. These could include:

1. You would like to increase the number of customers who can check out at a single register in one hour by 10%. What technology could you implement to make this possible? Develop a classroom simulation that shows how you can reach your target.
2. Your mom believes that fast-food drive-through lines take longer now than they did 2 years ago. Is she right? Develop a model of the processes, including any changes, and explain if she is correct. Develop some changes that could be implemented to speed up the time in line. Support your process designs with persuasive argument based on data.

These are just two ideas among many. Having a conversation with students after they have engaged in the grocery store problem could elicit any number of ideas for further exploration. Tapping into your students' imaginations and curiosity will make learning that much more meaningful for them and for you.

CHAPTER 7

How Constraints and Criteria Affect Design and Mathematical Modeling

The design of products and processes is driven by design criteria and constraints. *Criteria* (sometimes called requirements) are things that you want to incorporate into the design, and *constraints* are things you must have that cannot change. You are probably familiar with this type of approach even if you do not always think about it in these terms. Here are a few examples:

- Make a delicious meal for two for less than $10.
- Build a bookcase that will hold 100 hardbound books using found materials.
- Choose a book you can read during a 1-week vacation while leaving enough time for daily activities.

The criteria and constraints for these three examples are shown in Figure 7.1.

As you go about using and developing design activities, criteria and constraints can help you elicit the learning that you wish for them. In general, constraints can be used to narrow the possible solutions and target specific mathematical or scientific relationships in determining the efficacy of a designed artifact or process, while criteria allow for open-ended discovery. With reference to one of the examples above, there are many ways to make delicious meals for two, but far fewer ways to do it for less than $10. Changing the allowable cost has an impact on the allowable meals and changes how someone could approach the challenge. Lowering the cost to $2 would limit available ingredients and perhaps bring the use of vegetables, herbs, and spices to the fore. However, there is one additional aspect that needs to be considered. What constitutes "a delicious meal"? This is at the heart of assessing how a design can be measured against the criteria. In some cases, the rating by a judge or a panel of judges is appropriate, but in others an objective set of measures may be more appropriate.

In this chapter, we expand your understanding of the use of criteria and constraints to promote learning through example design and modeling

Figure 7.1. Example Criteria and Constraints

Criteria	Constraints
Make a delicious meal for two	It must cost less than $10
Build a bookcase that will hold 100 hardbound books	You must find the materials and not pay for them
Choose a book you can read in one week	You must have time for all your activities

activities. Some of these activities are strongly constrained, while others have very few constraints. While these activities can be connected to additional content standards, in these examples we provide only the most central standards that informed thinking about constraints. Our first example is of design-based learning with mathematical modeling and strong constraints that narrow the number of possible solutions.

HOW MANY MARBLES CAN YOU FIT IN A PIECE OF ALUMINUM FOIL BEFORE IT SINKS?

Mr. Napoleon teaches 6th-grade mathematics and wants his students to develop an understanding of the volume of right regular prisms (see Figure 7.2). To do this, he uses a strongly constrained design and modeling activity: Design and build a right regular prism from aluminum foil 5 cm in height that will hold exactly 500 grams before sinking in a tub of water.

For this activity, the area of the base must be 100 cm squared for every group's design; however, every group can make any shape they want for the base as long as it has a 100 cm area. A variety of prisms can be useful, and adding a criterion that the "coolest prism to meet the constraints will be judged to be the best" can be used to motivate students to find the surface area of more complicated shapes.

KEEPING A COLD DRINK COLD FOR LONGER (CO-CONSTRUCTING CONSTRAINTS)

Ms. Kowalski and Ms. Lundgren planned together in teaching 8th-grade science (see Figure 7.3). They wanted to develop a design activity that would help students better understand the nature of thermal energy and how it is transferred. After some discussion, they decided to develop an activity to make an artifact that would keep water cold for as long as possible. After trying some things out, they realized that the nature of thermal transport processes had a large impact on how quickly the water heated

Figure 7.2. Focal Standard in the Marble and Foil Activity

CCSSM Mathematics Content and Practice Standards

- 6.GA.2: Find the volume of a right rectangular prism with fractional edge lengths by packing it with unit cubes of the appropriate unit fraction edge lengths, and show that the volume is the same as would be found by multiplying the edge lengths of the prism. Apply the formulas $V = lwh$ and $V = Bh$ to find volumes of right rectangular prisms with fractional edge lengths in the context of solving real-world and mathematical problems.

and that students would need to make measurements to understand those processes. They would also ultimately need to add details about the volume of the water, the initial temperature of the water, and perhaps some constraints about the surface area of the water. They discussed how the surface area and volume of the water influence cooling rate and decided that they needed to use constraints to help set up the final design activity.

While they realized that they would have to develop a detailed list of criteria and constraints for student designs, they struggled a little with deciding the best way to address the criteria and constraints. Should they provide them to their students, or should they take some time to have the students experiment and explore first to see what they are able to discover on their own? They decided to try to co-construct the criteria and constraints with their students and discussed how they could go about getting their students started. They reflected on one of the disciplinary core ideas listed for this standard as they developed their lesson plan. They were encouraged that considering criteria and constraints was in line with these standards.

To prepare their students to consider the importance of constraints, they collected disposable cups from different local stores and made short videos with individual 8th-grade teachers and school administrators. Each teacher and administrator expressed in their video that they bought their iced drinks from a particular store because they felt they kept their drinks colder longer. They edited the video and presented it to the students, asking students to help teachers at their school figure out which cup really keeps iced drinks cooler longer.

Ms. Kowalski and Ms. Lundgren asked their students to make careful observations on the experiments they did and to keep track of everything they measured. They intentionally chose cups made with different materials and of different sizes. They made sure that they had different jugs of water, some with large quantities of cubed ice, some with crushed ice, and some with no ice. They also provided students with digital thermometers and stopwatches.

As students started working on filling cups and measuring temperature, Ms. Kowalski noticed that some of her student groups had used large

Figure 7.3. Focal Standard and Disciplinary Core Ideas in the Cold Drink Activity

NGSS Science Standards

- MS-PS3-3: Apply scientific ideas or principles to design, construct, and test a design of an object, tool, process, or system that either minimizes or maximizes thermal energy transfer.

NGSS Disciplinary Core Ideas

- ETS1A: The more precisely a design task's criteria and constraints can be defined, the more likely it is that the designed solution will be successful. Specification of constraints includes consideration of scientific principles and other relevant knowledge that is likely to limit possible solutions.
- PS3A: Temperature is a measure of the average kinetic energy of particles of matter. The relationship between the temperature and the total energy of a system depends on the types, states, and amounts of matter present.

amounts of ice in their cups and that the temperature was not changing at all with time. She asked to look at their data and asked her students why they thought they were seeing what they were seeing. One student asked if it had something to do with the ice. Ms. Kowalski was very pleased with this observation because it showed the beginning of making connections to PS3.A, which is one of the disciplinary core ideas for this activity.

Rather than confirm the student's observation, Ms. Kowalski said, "That's a really interesting observation. Can you think of a way to get a better understanding of whether ice has an impact on your measurements? How about you talk in your group for a few minutes and think about that." Ms. Kowalski continued checking in with other groups and noticed that some groups had filled every cup to the brim, while others had a similar volume in each cup and yet others had seemingly random amounts of water in each cup. She allowed them to keep working and did not mention her observations yet; however, she wanted to revisit this later and asked all of the groups to take pictures of their experimental setup with a group member's phone as part of recording their observations. As class was coming to an end, she reminded them to make sure they had all their data and that they would share their results after they had plotted them in the next class period.

In the next class period, after all the groups had plotted their data, Ms. Kowalski asked her students to share their data with her, and she combined the data from each of the groups one cup type at a time. After plotting the data for the first cup, Ms. Kowalski said, "This is very interesting [see Figure 7.4]. You got very different results for exactly the same type of cup. What do you think we can learn from this data? Discuss in your groups for three minutes and then we will discuss as a whole class."

Figure 7.4. Graph of Temperature Change in One Type of Cup Over Time

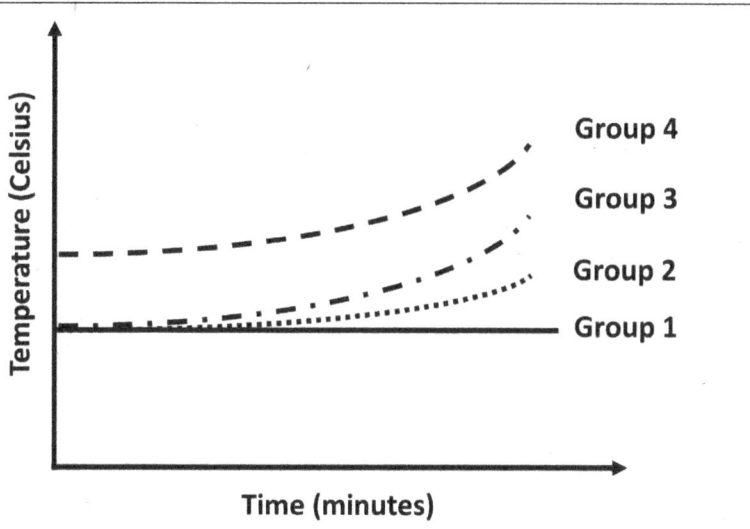

After waiting for a few minutes, Ms. Kowalski asked her students for their thoughts.

"We used ice in our cup and the temperature didn't change. We think that the ice kept the water in the cup at zero degrees Celsius" reported Group 1. The members of Group 4 reported that they used water with no ice and that it started at 5 degrees Celsius. Group 2 and Group 3 both used water that had a small amount of crushed ice that quickly melted, and initially they could not come up with a reason why their data were different other than discussing the way that they took the measurements, but both groups used the same method and the same equipment. Ms. Kowalski asked them if they had any other observations that they had not considered yet, and one student said, "Oh, what about our photographs?" "Great observation," said Ms. Kowalski. "Let's take a look." Groups 2 and 3 compared their photographs and realized that Group 3 had filled their cup to the brim, while Group 2's cup was only half full. "Oh, the more cold water there is in a cup, the longer it takes to heat up," said one of the students.

Ms. Kowalski was now in a good position to lead a class discussion on how to accurately compare the different cups by deciding on how cold the water was and how much water could be poured into the cups. This understanding that it takes more heat flowing into the cup to heat a larger volume of water and that the temperature of iced water stays the same until the ice melts was vital in taking valid measurements for comparison.

Figure 7.5. Criteria and Constraints in Cup Design Activity

Criteria

Water must be kept below 10 degrees for at least 45 minutes.

Constraints

1. The sealed cup must hold exactly 500 milliliters of water.
2. The water must initially be 0 degrees Celsius and not contain any ice.
3. The materials used must be nontoxic.
4. There must be space to place a thermometer in the water.

After completing a second round of measurements on the teachers' cups, Ms. Kowalski introduced a design activity with criteria and constraints.

> Design a sealed cup that will keep 500 milliliters of water that is initially at zero Celsius (but contains no ice cubes) below 10 degrees Celsius for at least 45 minutes. You may use any nontoxic materials but must leave space to place a thermometer into the water.

The constraints and criteria are summarized in Figure 7.5.

This activity is primarily an example of how constraints can be used to target specific learning outcomes. However, you could also modify the final design project with additional criteria if you wanted to emphasize different or additional topics. For example, you could add a constraint that the container must be transparent if you wanted students to better understand that radiation from the sun is one way for thermal energy to be transferred. Or you could add a criterion that the cup should be as light as possible if you wanted students to gain a better understanding of the difference in conduction of heat by different materials.

In this chapter, we shared how criteria and constraints can help focus on the specific learning goals you have for a design-based activity. In the examples, you can see how our teacher colleagues used constraints to narrow down the possible solutions and target specific mathematical or scientific relationships in determining the efficacy of designed artifacts. Notice also how the teachers varied how criteria and constraints were determined and shared. Doing so can alter the degree of open exploration in a task and, in turn, the cognitive demand. In the next chapter, we discuss how this and other approaches can scaffold tasks to support learner success.

CHAPTER 8

Scaffolding Student Learning in Design and Mathematical Modeling

It is early in the year, and students in Ms. Brandon's 7th-grade class are working on a task she has designed for them that draws on her experiences from a professional development opportunity she had taken part in over the summer. The focus of the professional development was on teaching mathematics and science through a design-based approach. She feels energized by what she has learned from the experience and wants to introduce her students to the design process early in the year so that they can continue to grow and develop their understanding over the coming weeks and months. She designed the following task to begin their exploration: Design a roller coaster on which it takes a marble 45 seconds to traverse. Students have card stock "track" components to select from, and Ms. Brandon tells them they can add additional materials of their choosing, as long as they can find them in the classroom. Students fold and cut card stock and determine how to stick pieces together to assemble their coaster designs. Their coasters must have a variety of components, and a minimum of three from the following options: straight track, sharp curve, wide curve, funnel, merge, and loop.

Ms. Brandon's students begin building their coasters and testing their models by determining how long it takes a marble to traverse the track. There is excited chatter in the classroom as her students begin. As students test their models, they record the results to determine how close they are to the target of 45 seconds. Using the results of each test, they go back and revise their roller coaster design. The marble always seems to finish too fast in initial designs, and students need to figure out how to make it stay on the coaster longer.

As her students test and revise, Ms. Brandon notices that one group has tested and revised their model multiple times. Students in some other groups have stopped working and have checked out. They are talking to each other about video games, the upcoming weekend, and so on. Others are continuing to work, but are starting to get frustrated, declaring that it is "too hard" or "impossible" to do. Ms. Brandon realizes that these groups are no longer working productively. There is definitely struggle, but it is anything but productive. She realizes that she must do something to address the situation, and do it quickly, before she loses her students' interest for good.

Situations like this can occur when working on tasks that are open-ended and new to learners, such as design-based integrated science, technology, engineering, and mathematics tasks. Many students may be used to having a different type of learning task, one in which they follow steps outlined by the teacher or by an instruction sheet to get an answer. However, that is not the nature of the approach we are advocating in this book. Rather, our approach involves students working on tasks in which they participate in a dynamic and iterative process, using divergent and convergent thinking to generate as many potential ideas and solutions as possible before deciding on and putting forth (and defending) their solution. This can be exciting and challenging for both teachers and students. Essential to the approach is providing the appropriate supports at the right time along the way. Otherwise, the approach may result in confusion and frustration, as it did initially for Ms. Brandon's students.

PRODUCTIVE STRUGGLE IN DESIGN-BASED LEARNING

As teachers, our driving goal is for our students to learn, to grow understanding, and, ultimately, to succeed. It can be difficult, even uncomfortable, to watch our students struggle as they are exploring and engaging with new concepts and skills. In some cases, our initial instinct may be to step in and "save" the student. Yet, if we do that, what valuable learning opportunities have we sacrificed? Cognitive scientists have advocated for the importance of struggle in learning, and research has demonstrated that struggle can play a key supportive role in the learning process (e.g., Dewey, 1929; Kapur & Bielaczyc, 2012; Piaget, 1960). The National Council of Teachers of Mathematics (NCTM, 2014) has described supporting students through productive struggle in learning as an effective teaching practice.

But what exactly is *productive struggle*? Definitions of productive struggle generally include three key ideas: (1) learners drawing on prior knowledge and experiences to solve new problems that do not have an obvious solution path, (2) learners persevering as they solve problems, and (3) learners working with important, relevant, and appropriate concepts and skills (Baker et al., 2020; Hiebert & Grouws, 2007). Instructional moves that can help create productive struggle in learning include engaging students in high-level tasks, focusing on justification and explanation, and valuing student agency and voice (e.g., Engle, 2006; Franke et al., 2015; NCTM, 2014; Stein et al., 2009; Warshauer, 2015a, 2015b; see Figure 8.1). Another key aspect is establishing a learning environment that

Figure 8.1. Instructional Moves to Help Create Productive Struggle

- Use high-level tasks
- Focus on justification and explanation
- Position students as contributors
- Value student agency and voice
- Normalize failure as expected and productive for learning

supports and normalizes the process of failing, making mistakes, and then persevering and learning from those mistakes. Struggle is a key part of that kind of learning. Important to understanding the learning that is possible through productive struggle is recognizing that the emphasis is on more than getting a correct answer. The focus is on making connections and understanding relationships. A design-based learning approach creates a prime environment for this type of learning.

We discussed in Chapter 3 that an affordance of design-based learning is that it creates the opportunity to engage students in productive struggle. One of our teacher participants, Mr. Potter, highlighted this idea:

> When you think of the classroom, a lot of teachers will line out each step that you take to get to a solution, but the engineering design we've been working on here doesn't want to use that hand-holding. You give us either the goal or a small set of criteria to follow and then we have to figure out the steps and in that part there's struggle, but it's productive because we either figure out steps that work or we don't figure out the steps that work, in which case we revise until we do figure out what works.

Another teacher, Ms. Vaughn, described how design tasks, by their nature, are open-ended and allow students to have more agency. She noticed that "letting kids have more ownership of whatever they are creating instead of giving them step-by-step, you know, do this, this, this, this; put that away and kind of let them discover on their own" resulted in more creative and engaged learners.

Both of these ideas align with instructional features and outcomes related to productive struggle. In our approach to design-based activities, learners draw on prior knowledge and experience to understand the issue or problem at hand and formulate ideas for potential solutions or parts of solutions. Learners persevere through building prototypes of their solution, testing them, and revising to improve them. The process of designing, redesigning, and evaluating solutions creates space for persevering through challenges. During the process, learners continue to draw on prior knowledge and experiences while making sense of important mathematics and science. All of these features fit with the practice of supporting productive struggle in learning.

SCAFFOLDING PRODUCTIVE STRUGGLE

Although research evidence may support the value of struggle, it can be difficult to carry out in ways that keep struggle productive. This is in part due to our cultural expectation that a teacher's job is to remove or prevent struggle. Another reason is that many educators did not experience productive struggle as learners ourselves, therefore, it may be difficult to envision what this would look like in a classroom. In fact, teachers may fear that the approach of supporting productive struggle may result in allowing students to remain stuck, possibly not even able to begin, or floundering about once they have reached an impasse. Teachers may have fears of students becoming frustrated and essentially giving up and checking out. However, this is not what it means to support productive struggle in learning. A key piece of *supporting and maintaining* productive struggle is scaffolding, in particular knowing when to scaffold and how to do so in an appropriate way. Scaffolding can be the difference between struggle that is just struggle and struggle that is productive (Barlow et al., 2018). Thus, knowing the appropriate what, when, how, and how much to scaffold is essential.

But how do you provide the right kind and amount of scaffolding, without giving too much, so that your students have agency and are persevering and struggling productively? Too much support risks compromising learner autonomy and deep learning as students simply follow directions, while too little support risks damaging their self-efficacy and perseverance as they struggle without making progress. What is the right level of support to maintain that balance? In truth, this level will vary depending on the task, the content, the context, the student, and even the day! And once you determine the appropriate level of scaffolding, what do you do to implement that support effectively? We turn next to discussing guidelines that can inform your reflective teaching decisions by drawing on both the research literature and our own experiences working with teachers as learners in professional development we have provided.

There are many ways to consider scaffolding to support and maintain productive struggle. For instance, Barlow and colleagues (2018) suggest strategies for scaffolding the task at the outset in order to provide access to students. These include activating prior knowledge, providing ample time up front to analyze the task before jumping in, and posing a simpler problem to initiate thinking. In addition to scaffolding the task, Townsend and colleagues (2018) share three areas of scaffolding as students work on tasks: (1) establishing and maintaining productive collaborative group work norms; (2) providing ongoing teacher support and motivation; and (3) establishing and maintaining a comfortable and supportive classroom culture. They suggest tasks that are embedded in real-world contexts that are relevant to students' lives. The focus of the tasks should be on looking for relationships and connections rather than a specific strategy or answer. Tasks should be structured so that they provide multiple entry points and draw on multiple strengths and perspectives. Student groups should be heterogeneous, allowing for multiple perspectives and skills that provide opportunities for divergent thinking. Each group member must be responsible for understanding and contributing. Ensuring that this expectation is met is the responsibility of the entire group.

As students engage with the task, your responsibility as teacher in this approach is to monitor group progress, ensuring that each member is understanding the task, strategies, suggested solutions, and so forth. This can be done through questioning each member about their progress. If it becomes evident that someone is not understanding, you can remind the group that they need to support one another and then facilitate a conversation among the group to address that incomplete understanding. In addition, you are responsible for looking for opportunities to assign status to students' ideas and contributions. For example, you can circulate, listening and observing as groups work. As a student offers an idea that is useful, insightful, productive, or novel, you should acknowledge it, elevating the contribution and assigning status. In particular, you should look for students who may not always be acknowledged in the classroom or by their peers. The intent is to clearly communicate high expectations for each student, to publicly acknowledge multiple ways of knowing, and to highlight the idea that diverse ways of thinking are strengths in reaching solutions to problems. You probably recognize these ideas and those in the previous paragraph from complex instruction, one of the evidence-based teaching practices discussed in Chapter 3. In addition, you should consistently emphasize that failure and struggle are part of the learning process. The key is to learn from each failure and use it to revise and improve.

Townsend and colleagues (2018) found that providing appropriate supports at the right time made students' struggles more productive. But when is the right time to provide support? Suppose that (a) you have done the work up front of selecting a task that is connected to students' lives and interests, provides multiple entry points, and engages them in meaningful content; (b) you have elicited students' relevant prior knowledge; (c) you have allowed students time to analyze and understand the task prior to working on it; and (d) you have perhaps even given a simpler task to begin. Now, as your students are working, how might you recognize moments when you need to step in and when it is most appropriate to let go?

Townsend and colleagues (2018) built on Vygotsky's (1978) notion of *zone of proximal development*, which includes things students can do with support that they could not do alone, and shared "indicators to analyze students' zones of productive struggle" (p. 218). These include several "look-fors": Were students persevering in the task? Were students asking questions of and providing support to peers? Did students give up, get off-task, and show indicators of frustration or express negative beliefs about themselves as learners?

Let us return to Ms. Brandon. She has noticed some of the look-fors that are indicators of a zone of unproductive struggle in some groups. In particular, she noticed that some students have checked out and are demonstrating frustration. What might you do as a teacher in this situation? For one, you could ask a question. The purpose of beginning this way is to allow students to use their knowledge and experience to consider ways to move past their impasse. If you tell them what to do, you are the one who is doing the thinking, likely drawing on your own understandings, which might not be as relevant to students as eliciting their understandings. As a result, you run the risk of losing the opportunity for students to make their own connections between existing and new knowledge and to deepen their learning. Therefore, ask a question or invite students to reflect: "Tell me what you have done so far"; "What seems to be the big issue?" or "What additional information might be helpful?"; and "How can we find that?" Often, this will be enough to get students back on track.

For example, in the roller-coaster activity, Ms. Brandon approached the group who had already tried multiple revisions and asked them, "What have you learned from your revisions? What does each element add to the time?" When students were not sure, Ms. Brandon asked them if there was a way that they could investigate and get an idea of how long it takes the marble to traverse a particular element on the track. After this, a student suggested that they could use their phones to record and time how long it takes for the marble to move along these different elements. Ms. Brandon acknowledged this suggestion and asked the other group members how they think they might be able to use that information to move forward. A student suggested that they use that information to determine what to add. Ms. Brandon agreed. Recognizing that the students had moved past their temporary impasse, she moved on after making a note to return to check on them in a bit.

In this scenario, Ms. Brandon recognized an issue of unproductive struggle and provided scaffolding for students to help them move past it. She asked questions to support their thinking about how to identify the specific issue and then asked additional questions to help them consider ways to investigate and redirect their solution. In doing so, Ms. Brandon assigned status to a suggested idea that she recognized would be productive.

In this example, Ms. Brandon responded to and provided scaffolding for an entire group. It is also necessary to recognize that you often need to offer different forms and amounts of scaffolding for different students. For example, if you have a student who struggles with organization, consider providing blank tables or charts that they could use to organize information. Then, check back in with the student periodically to see what they have done so far and ask what they are doing next. Helping students think through a plan of action and keep on track while carrying it out can be what they need to persevere in the task. If asking students to explain their process in writing, consider providing sentence stems to help students who have difficulty expressing their ideas better organize their thoughts and focus in on important ideas. If students struggle with writing, consider using audio or video to allow them to record their thoughts. Using audio or video in that way can provide you with better assessments of what students understand and how they think compared to forcing them to write it out. The key is to consider what supports specific students need to enable them to remain engaged in the work and the learning.

In addition to these ideas drawn from existing research literature, the following have been particularly successful for us and our teacher colleagues: (a) using formative assessment to address incomplete or inaccurate content knowledge that is impeding progress; and (b) giving students voice in design constraints, criteria, and methods of assessment. We next explore each of these strategies, drawing on examples from our work with teachers and their students.

STRATEGICALLY USING FORMATIVE ASSESSMENT DATA

Think about Ms. Brandon again. She notices that other students have taken up the idea of collecting data about the length of time it takes for each marble to traverse specific areas of track. However, once they have combined particular chains of elements, the timing is not working out as they had expected based on their collected data. She realizes that her students are not taking into consideration the effect of the marble's velocity as it exits one element and then enters another, compared to dropping the marble into an element at an arbitrary velocity or starting from rest and measuring the time to traverse that section. She also notes that students are using terms like *force, speed, energy, acceleration,* and *momentum* inconsistently and sometimes interchangeably.

She takes notes on the ideas and questions students are bringing up and decides to design and implement a mini-lesson on motion the next day. The goal of the lesson is to help students think about how mass and velocity affect potential and kinetic energy. In this lesson, students build and test ramps at various heights to launch a small marble to hit a target (see Figure 8.2). Students are to analyze and graph their results each time, accounting for distance, time, weight of the marble, and changing the slope of the ramp. After using the small marble, they next make predictions about how a large marble will act and whether it should be launched from the same height, lower, or higher on the ramp to hit the same target. They learn that the weight or size of the marble (mass) does not change its path to the target.

Figure 8.2. Design-a-Ramp Activity and Standards

You are to design a ramp that will launch a small marble into an open cup placed 10 cm below the launch point and 40 cm away from the launch point. Build the ramp out of the insulation provided (use the diagram below to help you). How high does the marble need to be placed on the ramp?

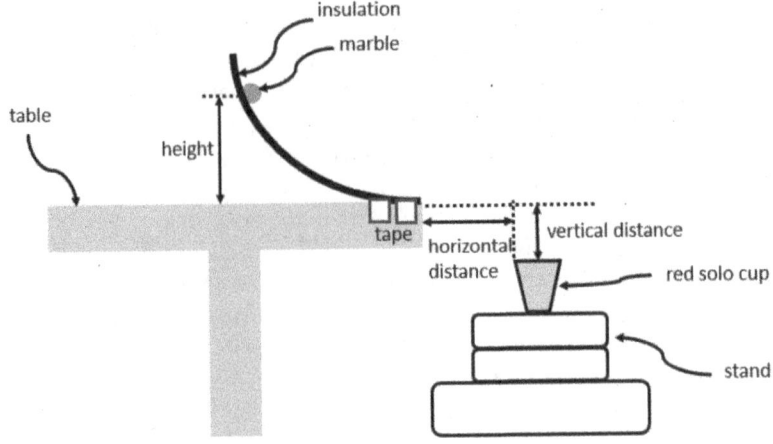

NGSS Science Standards
- MS-PS3-1: Construct and interpret graphical displays of data to describe the relationships of kinetic energy to the mass of an object and to the speed of an object.
- MS-PS3-2: Develop a model to describe that when the arrangement of objects interacting at a distance changes, different amounts of potential energy are stored in the system.

Next, students are told to cover the opening on the target with a thin napkin fixed in place with a rubber band and predict, using what they have learned so far, how high a large marble needs to be placed on the ramp in order to break the napkin and how that effects where the target needs to be located. Ms. Brandon encourages students to "find mathematics" to support their predictions. They test their predictions from the large marble and then make predictions before testing what will happen when they try it next with a small marble. It was important that they used the large marble first this time, and they learn, among other things, that the size of the marble does make a difference when the small marble does not break the napkin after dropping from the same height at which the large marble did break it. Testing to determine what height is needed to break the napkin with the small marble helps this comparison come through even more clearly.

Ms. Brandon has students discuss what their group has learned about mathematics and science each step of the way and before she engages in whole-class conversation about mass, velocity, potential energy, and kinetic energy. Finally, students write reflections on how what they have learned might be useful in thinking about their roller-coaster design. The activities Ms. Brandon designed in the mini-lesson encouraged students to distinguish between energy and force, and how velocity and mass are related to potential and kinetic energy. This lesson also incorporated mathematics and literacy practices related to measurement, graphing, mathematical modeling, and written reflections on learning.

Here, Ms. Brandon observed and listened to what students were doing and saying, taking notes of ideas that were being used or, perhaps, in need of some development; in other words, she was engaged in formative assessment. You probably also notice that she was engaged in both reflection-in-action as she was teaching and reflection-on-action later when she considered how her formative assessments could inform her next instructional moves. By doing so, she recognized that her students were revealing an underdeveloped area of understanding. In order to support them in moving forward, she decided to scaffold their progress by designing a fairly highly constrained or targeted design-based mini-lesson that allowed them to visit, revisit, and reflect on ideas that would be useful to their roller-coaster solutions.

GIVING STUDENTS VOICE IN DESIGN CONSTRAINTS, CRITERIA, AND METHODS OF ASSESSMENT

Ms. Felix, another teacher colleague, presented the following situation to her 6th-grade mathematics students:

> The planning committee for the school's Winter Festival fundraiser wants to change the cocoa recipe for the event. In the past, the organizers had been buying packets of prepackaged hot cocoa mix to sell. But people have not been buying it much; they say it is not very good. Because the school really needs to try to bring in the crowds this year to raise funds for the new track they want to build, the Winter Festival planning committee has tasked our class with trying to find a new recipe that will be a bestseller. Groups of students are to design and test a recipe or recipes, analyze the results, and adjust recipes until they are satisfied. Once groups have determined their recipe, they must determine the price per cup as well as how to scale up the recipe (and the cost) in order to serve the crowds at the Winter Festival.

There are several ways that Ms. Felix could approach determining design constraints and criteria. For one, she could set the constraints and criteria herself. For example, she could determine the constraints to be that there could only be five or fewer ingredients, the charge per cup must be around what they have usually charged at the Winter Festival ($1.25–$1.75), and they need to make a profit from their sales. Criteria may be that the ingredients should be easily available at local grocery stores, and it needs to be delicious! This removes some of the open-ended nature of the task so that students can focus on designing and redesigning their recipes, determining the cost, and scaling up. Further, Ms. Felix could provide students with a list of assessment methods by which their recipes will be evaluated to determine the winning recipe. These could include taste, cost and profit margins, availability of ingredients, and ease of recipe. She might create a rubric that assigns values to these categories based on importance. For example, taste and profit might be weighted twice as much as ease of recipe, based on the need to have a very successful fundraiser this year. Ms. Felix could also inform students that a panel of "experts" consisting of planning committee members, teachers, school staff, school administrators, and community members will judge the recipes and select the winning one using the rubric. Each student group would provide a sample of their cocoa as well as a presentation describing the recipe, cost, and how to scale it up.

However, an alternative is to have the students be involved in determining the constraints, criteria, and methods of assessment. This allows students more voice and agency in the process and encourages them to consider the issue from multiple perspectives. Ms. Felix can do this as a class by having students brainstorm all that could or should be taken into consideration. The class can then discuss which should be constraints and which criteria, and what methods of assessment should be used to judge winning recipes. The class may wish to collect additional data through informal surveys, Internet searches, price-checking, and so forth. All of this information would be recorded so that students could document how they determined their constraints and criteria. Ms. Felix could elicit ideas from the students about how the "best" recipe should be evaluated by gathering ideas (e.g., available ingredients, able to make big batches to sell, chocolate flavor), categorizing them into themes (e.g., taste, cost-profit margin, ease of recipe), and then letting students help determine the weight of each theme to create a rubric. Students can also be involved in determining who should evaluate the designs. Students can suggest who will judge the recipes by considering the various stakeholders, how the judging will take place, and what information each group must provide to the judges.

Varying elements such as who determines design constraints, design criteria, and methods of assessment can add to students' autonomy in their work on these tasks. The more that students have control over these issues, the more open-ended and cognitively demanding the task becomes. You, as their teacher, should engage in a process of scaffolding students to eventually let go and give them more say in determining these elements. As students are first introduced to design-based learning, it is best for you to take the lead in setting design constraints, criteria, and assessment. This allows students to become more familiar with exploring these tasks—which are open-ended, collaborative, and nonroutine—characteristics that may be unfamiliar compared to the types of tasks students are likely used to. You can focus on scaffolding students' access to the task, supporting productive group work, providing motivation and feedback, and providing necessary supports and scaffolding to meet the needs of individual students. As students become more comfortable with working in these ways, they can take on a greater role in determining constraints, criteria, and assessment tools. Perhaps you first provide constraints and ask students to determine criteria. Maybe you provide methods of assessment as the teacher and ask students to determine constraints and criteria. Over time, students can take on more and more as they are ready.

Our approach to integrated design-based learning aligns well with approaches to scaffolding learning so that productive struggle is facilitated and maintained. In this chapter, we discussed some guidelines for recognizing when support is needed to help students move past an impasse. We shared some specific strategies for offering that support to encourage perseverance, maintain learner autonomy, increase self-efficacy, and facilitate deep learning. As you begin your own journey of implementing design-based learning and take up these strategies, be sure to document how, when, why, and for whom they are working. You know your students best. The quality and quantity of support needed will vary by class, group, and individual student. Just as we ask students to design, test, and redesign, we need to do the same in our own teaching practice. Document what you have tried and the results of those efforts. You can use this information as you make plans for future instruction. It might even reveal a new or different strategy or combination of strategies that works well for your students!

CHAPTER 9

Design and Mathematical Modeling Across Content Areas and Grade Levels

In this chapter, we share examples of design-based learning from inclusive mathematics classrooms, inclusive science classrooms, special education classrooms, and across the 6th, 7th, and 8th grades, all of which include meaningful connections to mathematics, science, and literacy content standards and objectives. We discuss nuances of reflective teaching practice and how elements of our approach manifest differently based on a teacher's primary content area and grade level while focusing on appropriate content standards and objectives for those contexts. Our hope is that these detailed examples will help you better consider how to modify the approach we share in this book so that it works well in your own classroom.

GINGERBREAD HOUSE LESSON
(6TH-GRADE MATHEMATICS CLASS)

The gingerbread house lesson was created by a 6th-grade mathematics teacher, Ms. Charles. She was looking for a better way to have her students explore and find areas of two- and three-dimensional shapes. She also wanted a more authentic way than her textbook provided for her students to work with nets, which are two-dimensional representations of three-dimensional shapes. She designed the gingerbread house activity with these goals in mind (see Figure 9.1).

She introduced the lesson after students had been studying weather and the damage large weather events do to the landscape and environmental habitats as well as to homes and buildings. Students became interested in how to create structures that could withstand weather-related disasters. The task that Ms. Charles presented to her students was to work in groups to design on graph paper a house that has a minimum of 15 square inches, a solid foundation, a functional roof, and at least one entrance. In addition, they needed to try to build the house as tall as possible to preserve views of

Figure 9.1. Standards Addressed in the Gingerbread House Lesson

CCSSM Mathematics Standards

- 6.GA.1: Find the area of right triangles, other triangles, special quadrilaterals, and polygons by composing into rectangles or decomposing into triangles and other shapes; apply these techniques in the context of solving real-world and mathematical problems.
- 6.GA.4: Represent three-dimensional figures using nets made up of rectangles and triangles, and use the nets to find the surface area of these figures. Apply these techniques in the context of solving real-world and mathematical problems.

NGSS Science Standards

- MS-ETS1-2: Define the criteria and constraints of a design problem with sufficient precision to ensure a successful solution, taking into account relevant scientific principles and potential impacts on people and the natural environment that may limit possible solutions.
- MS-ETS1-3: Define the criteria and constraints of a design problem with sufficient precision to ensure a successful solution, taking into account relevant scientific principles and potential impacts on people and the natural environment that may limit possible solutions.
- MS-ETS1-4: Develop a model to generate data for iterative testing and modification of a proposed object, tool, or process such that an optimal design can be achieved.

CCSS-ELA Standards

- W.6.1: Write arguments to support claims with clear reasons and relevant evidence.

the surrounding landscape. Students had prior experience using formulas to calculate areas of rectangles and surface areas.

On the first day of the lesson, Ms. Charles introduced the task and assigned groups, and students got to work on their graph paper designs. Ms. Charles considered a range of factors and skills, combining them to create *heterogenous groups* with diverse skills to facilitate group success. For this task, she considered students' social skills and status, spatial visualization skills, artistic and design skills, organizational skills, and mathematical computational skills. Strong spatial visualization skills were needed in the task as students envisioned their three-dimensional house and how it could be represented in a two-dimensional net. Strong artistic and design skills supported students as they considered the aesthetics of their house. Strong computational skills were needed to accurately determine calculations of area. Finally, strong organizational and social skills contributed to the solution process by keeping group members on task and on board throughout. Ms. Charles composed groups so that members brought complementary skills to the task, allowing all to contribute to its solution. As groups worked, Ms. Charles walked around, observing and asking questions. She recorded notes on how groups were working together, the ideas that were brought up and discussed (or dismissed), and students' progress on their task.

Students recorded their measurements and submitted their designs to Ms. Charles at the end of the class period so that she could look them over and determine each group's progress. As she looked over the work of two groups, she realized that they were incorrectly applying to all shapes the formula for calculating the area of a rectangle. They had simply adapted it by multiplying a combination of two or more dimensions of a nonrectangular shape. She decided to begin the next day's class with a mini-lesson designed to reinforce the meaning of area so that students could use their conceptual understanding to then find the area of nonrectangular shapes.

The next day, Ms. Charles gave each pair of students 12 color tiles and asked them to create a shape with an area of 12 square units. The caveat was that each tile must touch another tile completely on a side. She asked her students to then determine the length of the sides of their shape. Then she had two student pairs present their shapes. One was a rectangle and one was in the shape of an L. She used a whole-class discussion to draw attention to the ideas about area that she wanted to focus on, asking questions such as, "How do we know that both of these have an area of 12 square units? Would the formula $l \times w = A$ work for the L-shape? Why or why not? What if I created a different shape (showed one that was like a stair-step)? What is the area? Would the formula work?" From this conversation, the class was able to agree that area is the space inside a shape and that the formula, $l \times w = A$, only applies to finding the area of rectangles. For other shapes, they could count the square units, break it up into rectangles (or other shapes for which they knew a formula), or some combination.

Following this, she returned each group's designs and encouraged them to double-check their calculations. She walked around and made sure to check in with groups that had demonstrated misconceptions the previous day. After they had checked their calculations, groups were to build their structures using materials Ms. Charles provided, including one box of graham crackers, a piece of cardboard to use as the base of the house, a tub of icing to secure the structure, and candy to decorate it. Their goal was to stay true to the original design as much as possible.

On the following day, students were asked to measure and calculate the area of the base and the height of the structure. They were to reflect on how the dimensions of the original design compared to that of the actual structure and what constraints affected their design. Following this, each group's structure would undergo one stress test using a fan to simulate a high-wind weather event and another stress test shaking their structure on an unstable desk to simulate an earthquake. Students would collect data on the performance of their structure, and engage in group conversations to consider what worked well, what did not work well, and why.

On the final day of the lesson, students looked over the data they had collected the day before and evaluated the performance of their structure. They then devised a plan for redesigning their structure to improve its performance. As an additional component, Ms. Charles told students that they were going to create this design to appeal to a prospective buyer. The buyer had the following criteria in ranked order: (1) a sturdy structure that could withstand weather conditions; (2) at least three stories; (3) below budget; and (4) aesthetically pleasing. She provided them with an itemized cost for their materials and the buyer's budget range. Using this, students calculated a sales price that allowed them to make a profit. Finally, students were to create a presentation using a medium of their choice (e.g., video, slides, written document, poster), describing their methods of assessment, their redesign, and responses to the following key questions:

> How did the design of the base affect the structure of your gingerbread house? What variable(s) affected the structure of your gingerbread house? What variables did you change in the redesign of your gingerbread house? Why? What additional factors did you consider as you created your redesigned house to appeal to the prospective buyer? Why were these important? How did you determine your sales price?

Ms. Charles assessed each group's work on the accuracy of their calculations, sturdiness and functionality of the structure, aesthetics of the structure, analysis of the structure's performance, redesign ideas, strategies used to appeal to their buyer, determination of the sales price, and reasoning for their decisions as shared in the presentation. She also assessed each student on their participation and their group citizenship (collaborated, contributed, and respected peers).

Ms. Charles addressed several mathematics and science standards in her lesson. In mathematics, she focused on content that was not entirely new to her students; they had worked with area formulas before. However, she did not feel that these ideas had been explored in a way that connected to real-world contexts.

Although Ms. Charles taught mathematics, her lesson addressed several science standards. It also built on ideas that students had been discussing in science and social studies. Ms. Charles had listened to students and consulted with other 6th-grade teachers to come up with a context that would allow her to address mathematics content in a design-based lesson that integrated other content areas while tapping into her students' interests. She implemented elements of complex instruction by assigning a group-worthy task and creating heterogeneous groups to work on it. She considered the various skills and understandings that would be valuable in supporting success and used that to assign students to their groups.

Ms. Charles included mathematical modeling in several areas. First, students used a two-dimensional visual model to determine their initial design. They had to make decisions about the best approach for meeting the criteria of the task, including what factors would be necessary for a stable structure. Next, they tested and analyzed their solutions and used this to consider how to redesign their solution to meet a realistic situation. She added constraints and criteria by introducing the buyer's perspective. Here, students needed to make decisions about what factors they should highlight and what to minimize (and to what degree) to make a structure that would appeal to the buyer. They also had to determine how to set the sales price in order to fit within the buyer's budget but still make a profit.

She engaged in reflective teaching by looking at each group's work and identifying an area of content understanding that needed to be addressed. She then provided some targeted instruction in this area to ensure that all students had access to the knowledge they needed to continue with the task. Each day, she looked over the notes she took from her observations, considering whether she needed to adjust anything. Because her students were still new to these types of tasks, she provided the constraints and criteria used to assess their work. At this point, she wanted students to focus on making sense of the context, creating their designs, analyzing their designs, and working collaboratively. She planned to continue to scaffold and provide space for additional student voice and agency in making these decisions as her students became more comfortable with design-based tasks.

PREDATOR/PREY LESSON (8TH-GRADE SCIENCE CLASS)

The predator/prey lesson was created by a grade-level team of 8th-grade teachers, including a mathematics teacher, a science teacher, and a special education teacher. Mr. Potter, the special education teacher, was the lead on the lesson, and some components occurred during his class while others occurred during his colleagues' classes. Their school was in a rural area, and many of their students had experiences with hunting. This meant that many students were familiar with the fact that limits for each season were determined, in part, by wildlife populations. Mr. Potter and his colleagues designed a lesson so that students could use this background understanding to explore how genetic variations and natural selection influence populations and population traits over time (see Figure 9.2). Prior to this, students had some experience developing Punnett squares, which are tables to determine possible outcomes for genetic crosses between individuals

Figure 9.2. Standards Addressed in the Predator/Prey Lesson

NGSS Science Standards

- 3-LS4-2: Use evidence to construct an explanation for how the variations in characteristics among individuals of the same species may provide advantages in surviving, finding mates, and reproducing.
- MS-LS4-4: Construct an explanation based on evidence that describes how genetic variations of traits in a population increase some individuals' probability of surviving and reproducing in a specific environment.
- MS-ETS1-4: Develop a model to generate data for iterative testing and modification of a proposed object, tool, or process such that an optimal design can be achieved.

CCSS-M Mathematics Standards

- 6.SP.5: Summarize numerical data sets in relation to their context.

CCSS-ELA Standards

W.8.1: Write arguments to support claims with clear reasoning and relevant evidence.

---

with known genotypes, for one and two traits, assuming they operate by independent assortment.

On the first day of instruction, students participated in several activities designed to explore bird traits and species of bird predators. First, students watched a video that gave an overview of these topics. Mr. Potter then asked students to make a prediction about how different traits might affect a bird's survival. Following this discussion, Mr. Potter gave each student a card that would indicate what traits their bird had (large or small, green or brown). Students played a game of freeze tag in which they were "birds" and tried to get past the "Mantis" (the predator). Students were told that birds who are small may run (others had to walk) and that those who were brown were not allowed to be tagged. Birds who were successful in getting past the Mantis could then find a partner with whom to pair their alleles. Students who found a partner then predicted the probability of traits being passed down to their offspring. Birds who were tagged by the Mantis had to sit out a round. Mr. Potter had a number of sticks with the possible genotypes written on them that a student who had been tagged out could draw from to simulate random inheritance. After doing so, the student was reborn as a new bird and could re-enter the game.

Students played several rounds, collecting data on the number of birds remaining and their traits after each round. As they played, students asked what happened if no birds had been tagged and were sitting out, meaning that no new birds were born. Mr. Potter used this as an opportunity to discuss population density and overcrowding. Students also asked why small birds could run and brown birds could not be tagged. This allowed the opportunity for a discussion about how mutations may occur that are advantageous to the survival of the species. Mr. Potter took notes during the discussion, which he shared with his colleagues, so that they could analyze the ideas students were sharing. They used these data to design discussions in later classes that reviewed and reinforced ideas related to relevant content standards and addressed students' ideas that were not quite on track.

The participating teachers began instruction in each class with a summary of some of the big ideas students had discussed the previous day. They also used questions designed to challenge and redirect ideas that students had shared that were off track. Students worked in groups to use the data collected from the freeze-tag game to create a graphic display of their choosing. Teachers encouraged students to look at trends related to the traits of the birds that survived and those that did not. As students worked, teachers walked around and asked questions to assess students' understanding of the essential ideas and goals of the lesson. After groups completed their graphs, they presented them to the class.

Mr. Potter used group presentations to launch a whole-class discussion about what information they could derive from the graphs. Some students had used bar graphs, others, a pie chart; some students had used the actual counts of the birds, and others had converted the quantities to percentages. This allowed the class to discuss the advantages and disadvantages of these decisions and what information was made visible (or not) by making each choice. It also was a time for students to ask questions and for Mr. Potter to facilitate discussion to try to address incomplete understandings. Following this, each student used the information gathered and displayed in the graphs as well as from the discussion to write a journal response to two prompts: (a) What might affect the survival and reproduction of a bird? and (b) Why might a certain trait increase or decrease within a population over time?

Mr. Potter and his co-teachers used a rubric to evaluate students' understanding of traits, how they are passed down, and how they can affect the survival of a species. The data the teachers used included notes collected during students' work on their group graphs, ideas brought up in the discussion, and students' responses to the prompts. The predator/prey lesson addressed several standards. It connected these ideas to a context that connected to students' lives and to which they could relate.

Mr. Potter and his co-teachers implemented elements of complex instruction by assigning a group-worthy task, creating heterogeneous groups to work on it, and assigning status to student contributions. Creating heterogenous groups of students with different skill sets ensured that all students in each group had something to contribute to the solution. Like Ms. Charles, Mr. Potter and his co-teachers considered the various skills and understandings that would be valuable in supporting success and used that to assign students to groups so that individual students' strengths complemented one another. They assigned groups that included students who had demonstrated understanding in previous work exploring Punnett squares for one and two traits, students who had experience and interest in wildlife and wildlife populations, students who were good at organizing data and selecting and creating appropriate data displays, and students who had strong skills in being able to read a graph and make inferences from it. By taking up students' ideas in the discussion, the teachers assigned status to their students' thinking and reinforced their identities as students with skill in science and mathematics.

The lesson engaged students in mathematical modeling to create a descriptive model using the data they collected. The CCSSM describes a *descriptive model* as one that "simply describes the phenomena or summarizes them in a compact form" (NGA and CCSSO, 2010, High School Modeling, para 6). In this lesson, students drew on the data they collected and graphs they created to build a descriptive model that described factors influencing the survival of a bird species and how traits may play a role in that process.

Mr. Potter and his co-teachers engaged in reflection-on-action by collecting and analyzing the ideas students shared during discussions. They then used that information to begin the next day by reinforcing essential ideas and addressing potential misunderstandings. They also engaged in reflection-in-action by asking questions during group work, deciding when to provide additional scaffolds and supports to ensure all students had access to the mathematics and science. The lesson was designed for a science class; therefore, the main content focus was science. However, students used mathematics, specifically data analysis and data display, to make sense of the ideas they were learning, discussing how the use of different displays might provide varying insights into the data.

SKATER RAMP LESSON
(7TH-GRADE MATHEMATICS AND SCIENCE CLASS)

The skater ramp lesson was developed and taught by a team comprising a special education mathematics and science teacher, Ms. Clover, and a mathematics teacher, Mr. Polly (see Figure 9.3). The lesson was designed for 7th grade and had some elements that were addressed in students' science class and other elements that were explored during mathematics class. The essential question explored in the lesson was, How would you describe the relationship of kinetic energy with mass and velocity (speed)? Many of the 7th-grade students were interested in skateboarding, and the town had just completed construction of a skateboard park. The teachers used this context for their lesson.

Figure 9.3. Standards Addressed in the Skater Ramp Lesson

NGSS Science Standards

- MS-PS2-1: Apply Newton's Third Law to design a solution to a problem involving the motion of two colliding objects.
- MS-PS3-1: Construct and interpret graphical displays of data to describe the relationships of kinetic energy to the mass of an object and to the speed of an object.
- MS-PS3-2: Develop a model to describe that when the arrangement of objects interacting at a distance changes, different amounts of potential energy are stored in the system.

CCSS-M Mathematics Standards

- 7.RP.A.1: Compute unit rates associated with ratios of fractions, including rations of lengths, areas, and other quantities measured in like or different units.
- 7.RP.A.2: Recognize and represent proportional relationships between quantities.

CCSS-ELA Standards

- W.7.1: Write arguments to support claims with clear reasoning and relevant evidence.

Ms. Clover started by asking her students, "What is momentum?" She encouraged them to think of a time when they may have heard this word and what they remembered about it. She recorded their ideas on a large Post-it paper. She then showed students a short YouTube video (https://www.youtube.com/watch?v=vnM_RmkRUtU) that showed a child riding a scooter down a ramp and landing in a foam pit, as an example of momentum. After watching the video, she led her students in a discussion about what they had seen, what they now understood about momentum, and how they would craft a definition. Next, Ms. Clover presented her students with a problem: How can you design a new ramp for the park that will get a skater to land in a foam pit?

Once Ms. Clover introduced the problem, the class discussed and generated a list of ideas, variables, and factors they would need to consider in solving it. Ms. Clover then told the students that they were going to design models of the ramp and use a marble and a plastic cup to simulate the skater and foam pit, respectively. Before the next class, she and Mr. Polly looked over the ideas that students had generated and realized that there were some additional ideas they would need to know in order to work on the task. She knew that some of her students identified as needing special education services struggled with organizational skills. Therefore, she and Mr. Polly decided to help them create a method of keeping track of ideas important to the task.

The next day, Ms. Clover began by telling students they would need to explore a few additional ideas, including mass, velocity, speed, gravity, kinetic energy, and potential energy. The students created a foldable, which is a three-dimensional graphic organizer, for these terms, including definitions of, formulas for and examples of each. Ms. Clover also had students use the formulas for the terms on a few examples to support any issues with computation. At the end of the session, students worked in small groups to generate a list of materials they would need to build and test their ramps. Finally, the class shared their ideas in a whole-class discussion and made a final agreed-upon list of materials.

On the third day, students began designing their ramp. Ms. Clover placed her students in heterogeneous groups and asked them to hypothesize about how much potential and kinetic energy a ramp would need to launch the marble into a cup 30 cm away from the end of the ramp. As groups discussed their hypotheses, Ms. Clover walked around listening to their conversations. She asked them to explain their reasoning for their hypotheses and took notes about their thinking. Students then created their ramps. They calculated the mass of their marble, the height of their ramp, the total length of the track, and the time it took to travel from the top of the ramp to the end of the ramp, recording their data on a lab sheet that Ms. Clover provided to help with organization. Students then used these values to calculate the amount of potential and kinetic energy their ramp had and compared it to their original predictions, recording these calculations and their observations about the success of their ramp on their lab sheet. They then discussed revisions to their hypothesis and redesigned their ramps. They recorded their revisions and their reasoning for them on their lab sheet before repeating the steps above.

In their mathematics class, Mr. Polly had students use data they had gathered from their work on the ramp to calculate unit rates and represent proportional and nonproportional relationships with tables, graphs, and equations, using the formulas for finding kinetic and potential energy. Mr. Polly led students in a whole-class discussion of what they had learned and generated ideas of what variables should go into a generalized model for creating a ramp that would work for a foam pit any distance away. Mr. Polly and Ms. Clover had collaborated to review the formative assessment data from the science lesson, so he used this to plan some questions that would support students in considering what would be important. Students then worked in pairs to create their models and test them by modifying the distance of their cups. They either wrote a journal entry or recorded a video journal entry of their results and explained what they thought worked and why.

Ms. Clover assessed her students based on their conceptual understanding of kinetic and potential energy, their fluency and accuracy in their calculations, the design and redesign of their ramps, their reasoning and justification for their original and revised hypotheses, and their collaboration. Mr. Polly assessed students' calculations, their results, and their explanations for them.

The skater ramp lesson addressed several standards. The teachers used local contexts and students' interests to create a lesson for exploring these ideas. The lesson design had students working in both their mathematics and science classes. Therefore, the teachers could focus a bit more attention on each subject in their respective classes while being able to draw on what students had learned in one context to help understand the other, creating a more-integrated science and mathematics learning experience.

Ms. Clover implemented elements of complex instruction by assigning a group-worthy task and creating heterogeneous groups to work on it. As in other lessons, she considered the various skills and understandings that would be valuable in supporting success and used that to assign students to their groups. She considered students who had experience with skateboarding, those who were good at building and designing, students who had a good understanding of kinetic and potential energy, and students who were skilled at using formulas and carrying out computations. Mr. Polly used information gathered from reviewing the formative assessment data collected in Ms. Clover's class to determine his groups for activities in his class. He considered who had a good understanding of the science components as well as who demonstrated proficiency in mathematical calculations and writing equations.

The lesson provided opportunities for students to use mathematics to model the situation as they designed specific models for their ramps (Groshong, 2016). They drew on ideas learned in their science class to then develop a general model that would work in multiple situations. The teachers engaged in reflective teaching by using data collected from students to determine areas needing additional instruction to support students' progress on the task. They considered their students' needs and strengths to form groups and provide scaffolding through organizational tools.

ADAPTING DESIGN-BASED LEARNING
ACTIVITIES TO YOUR STUDENTS

We have provided a few examples of how our teacher colleagues implemented our integrated design-based learning approach in their teaching. In each case, you can see elements of our approach. Each lesson utilized a design-based task that was situated in the real world and drew upon students' interests and prior knowledge. These tasks were group worthy, and the teachers used thoughtful, heterogeneous groupings to support student progress and honor different ways of knowing and thinking. Each lesson implemented various forms of scaffolding and support. Teachers engaged in reflection-in-action by observing and listening to students, asking questions to scaffold or extend thinking. They also engaged in reflection-on-action as they examined formative assessment data and adjusted instruction to support student progress. Finally, they addressed important mathematics, science, and literacy standards in an integrated way.

While the lessons all had these common elements, there were variations in the lessons as well. As you may notice, in each example, the teacher designed instruction in a way that met the needs and interests of their students. In the gingerbread house lesson, Ms. Charles drew on interests she knew students had developed from their study of weather incidents in their science class. Even though she taught mathematics, she was able to address mathematics, science, and literacy standards in her lesson. In the predator/prey lesson, Mr. Potter and his colleagues leveraged their students' interests in and experiences with wildlife and hunting to design a lesson focused on science standards for their grade. Even though the lesson was science focused, they utilized mathematical modeling by encouraging students to collect and use data and make inferences from that data to develop a model and literacy practices to explain what influences the birds' survival. Finally, the skater-ramp lesson was implemented across both science and mathematics classes. The teachers here reflected on the lesson and student progress together. They worked collaboratively to discuss students' understanding using formative assessment data. This allowed them to make decisions about necessary supports and scaffolds, areas of underdeveloped understanding, and how to use students' strengths to form productive groups.

Another important note is that our teacher colleagues often took tasks that they had seen in other spaces and adapted them to fit the learning needs and interests of their students. For example, you might recognize that the skater-ramp task is closely related to the Design-a-Ramp mini-lesson described in Chapter 8. Ms. Brandon used that mini-lesson to scaffold her student's understanding of physics. We had, in fact, created and used a version of this activity in our professional development with our teacher colleagues, and several of them adapted the activity for their own classrooms in different ways. This serves as an example of how existing activities can be revised to fit your own teaching and learning context as needed, an issue we return to in more detail in Chapter 12. In this example, teachers drew on what they knew of their students' interests to revise a design-based activity to be embedded in a more real-world context that the teachers were confident was familiar to their students.

What we hope you take away from these examples is that knowing your students is key to succeeding with this approach. While we offer some ideas and recommendations, we recognize that you know your students as learners better than anyone. We hope that knowledge, drawing on these examples, and considering the elements of our approach provide a launching point for you to create tasks and lessons that spark the knowledge, excitement, curiosity, and creativity of *your* students!

CHAPTER 10

Design and Mathematical Modeling Across Instructional Modalities

In Chapter 9, we described a module developed by Ms. Clover and Mr. Perry around designing a skater ramp. It worked well in the classroom, but could it be modified for use with students who were studying from home? What if the module were delivered during a time when instructional modalities changed unexpectedly for everyone? Changing modalities has been a significant challenge over the last few years and can be particularly challenging for many hands-on and collaborative activities.

Some of the instruction you provide is likely to occur outside your classroom. In this chapter, we provide two examples of how this might be done using the approach you learned about in Part I (see Figure 3.1). On the one hand, you might do this for reasons as simple as wanting to give students additional experience or supplemental instruction beyond the school day. Mr. Shaker asking his students to find someone at home to help them think through how to make the perfect hard-boiled egg, as described in Chapter 1, is a simple version of this. On the other hand, more-extensive instructional support is needed when students have extended absences due to health or other issues and when schools close or go virtual due to extreme weather events, other disasters, or even global pandemics.

During the global pandemic, there were many changes that created an unplanned experiment in teaching and learning (Usak et al., 2020). Those changes included a reduction or elimination of in-person instruction, less exposure to new content, and little to no peer interaction for students. Teachers and students struggled with new technologies and approaches to teaching and learning, with those approaches varying tremendously across classes, schools, school systems, and states depending on their reaction to the pandemic. On top of this, reduced or lost employment, food insecurity, and other basic concerns created substantial stress in many students' (and teachers') homes, homes that in many cases now became the place where teaching and learning occurred. Lack of technology access and Internet connectivity at home became critical barriers in some contexts. Those barriers and stresses were most likely to impact those students (and teachers) living in lower-income urban and rural areas—and more severely than they impacted many of their peers.

TAKING DESIGN-BASED LEARNING ONLINE

While there were some attempts to implement project-based learning online, students and teachers struggled to implement it effectively (Miller et al., 2021). As you have seen throughout this book, some of the strengths of design-based learning are related to students interacting with peers in groups and struggling together while receiving instructional support and scaffolding from teachers. These can be challenging without in-person instruction. When not all students have access to the same resources in their homes, how can equitable education occur outside the school building? How do we redesign existing activities when instructional modality has to change? How can we design rich activities that can be used across modalities without damaging equity and broadening learning gaps?

Starting with the design process and reflection can be powerful (see Figure 3.4). Remember that the design process is iterative and dynamic, so you will not move smoothly from one stage to the next and end with a "finished" design. Rather, you will focus on considerations in one stage and then reflect on which stage makes the most sense to engage in next. Communicating with your professional network to share challenges and successes is helpful throughout every stage.

Begin with the Empathize design stage and consider what you know about your learners, additional needs your learners have when learning at home, and what additional pressures may be at play due to whatever caused instruction to shift modality. How do these impact the ways in which your instruction should be implemented to effectively scaffold learning and maintain productive struggle? This will eventually lead you to prototype, test, and redesign different kinds of activities, technologies, and scaffolding for learners. When learners are at a distance, it is important to build in formative assessment to replace your ability to assess them informally as you circulate through the classroom watching them work. More or different scaffolding is often needed to keep them engaged and struggling productively. As you work in the Define and Ideate stages of the design process, constraints and criteria will be important to consider differently for instruction that does not occur in person.

In Chapter 7, we focused on how constraints and criteria affect design and mathematical modeling. Remember that constraints are parts of the context for design that cannot be changed. Criteria are things we would like our design to include, things that would be nice to have but are not strictly required. Teachers (and sometimes students) can determine what constraints and criteria to include. Constraints such as the following are important to consider when designing for instruction at a distance:

- Teachers cannot directly observe students as they go about their activities.
- Students cannot interact with one another in the same physical space, but some students may be able to interact through virtual platforms.
- Some students may not be able to purchase supplies.
- Some students may not have reliable or high-speed Internet connectivity.
- Some students may have limited access to indoor space to build or work.

It is important to note at this point that constraints impacting "some students" have a troubling potential to introduce what we consider to be *unacceptable inequities*. If some of your students cannot purchase supplies, do not have reliable Internet service, or do not have a safe place to build and work when learning activities require these things, then those students will not have access to learning opportunities that their peers engage in. Notice that those students most likely to be negatively impacted in this way are also likely to have fewer economic and social resources within their families. Deep reflection on what you know about your students in the context of the equity framework we introduced in Chapter 3 is important to avoid reinforcing and broadening such social inequities (see Figure 3.2). Your design-based learning activities need to be developed in ways that avoid disadvantaging any student.

In addition to the constraints described above, we also suggest some criteria you should address in developing design-based learning activities to be implemented at a distance. These could include:

- We would like students to interact with their peers.
- We would like to provide students with appropriate scaffolding adjusted to individual student needs.
- We would like to utilize technologies students are already familiar with.

In some teaching contexts it may be possible to have students use video conferencing software to interact virtually; however, for some students their only access to the Internet may be through low-bandwidth devices or an adult's borrowed cell phone. Bandwidth and the need for specialized logins, such as if you decide to use multiple different apps or technology platforms, can be a problem for some students when accessing content.

For this reason, *we suggest learning about social media your students use themselves and considering how to leverage those platforms to provide scaffolding.*

These constraints and criteria highlight the challenging nature of planning across modalities, but design thinking can be used effectively to design learning activities for implementation at a distance. We next provide two examples and then connect those examples to each core component, affordance, and support in our approach (see Figure 3.1).

EXAMPLES OF DESIGN-BASED LEARNING AT A DISTANCE

The two examples we provide next differ in the resources needed to do the activity. In the first example, Ms. Patel could rely on technology resources provided to all students in her school. In the second example, Ms. Cartwright had to shift modality when her school went virtual and her students did not have reliable technology or Internet access. You will find a third example described in Chapter 12 (see Figure 12.3). In each case, our teacher colleagues drew on what they knew of our approach as they designed and implemented design-based learning activities at a distance.

Designing a Meal Schedule

Ms. Patel taught 7th-grade mathematics during the pandemic in a school district where students were provided computers at home through the school and all her students had reliable access to the Internet. She thought the Empathize and Define stages of design thinking might be particularly relevant places to have students gather information on potential users that would involve their caregivers and be relevant to their home.

Ms. Patel was working with her students on algebraic relationships and graphing data. She wanted to give them a meaningful design-based learning activity while they were studying away from the classroom. After some consideration, she developed an activity that she hoped would allow them to work with their caregivers in authentic ways while giving some opportunity for customization.

Ms. Patel developed a short survey for students to ask their caregivers about their priorities related to preparing food, including items such as vitamins and minerals, energy, time to prepare, and cost. Students collected data and collaborated online in small groups on a spreadsheet to create bar charts of the items. Students then interviewed their caregivers to help them create a problem statement that expressed their priorities. Ms. Patel reviewed initial drafts of students' problem statements and provided feedback until those problem statements were in a form like the following example: "Our family needs meals that can be prepared in 20 minutes or less that provide good nutrition, are low in sodium, and cost less than $20 for a family of four." She then had students work in their groups to develop criteria and constraints for the meals and to research the nutritional content of different foods. Students were then free to explore designing meals however they saw fit.

She allowed students to struggle with the final design of the meals and with developing the problem statements, but she took a more structured approach early on by developing the topic and the survey herself. Students had more autonomy in the interview and how they chose to explore designing meals. This can be more challenging in other settings where your students may not have access to reliable online collaboration platforms, but, as the next example highlights, there are ways to work around that.

Designing a Bird-Feeding Program

Ms. Cartwright taught 6th-grade science in a rural school district where there was very little access to broadband in any of her students' homes and many students had limited access to technology. The school coordinated pickup and drop-off of supplies once a week during the pandemic. Ms. Cartwright knew that this would allow her the opportunity to send supplies home with her students for design activities. She developed a 3-week learning activity centered around designing a bird-feeding program that initially targeted a science standard focused on maintaining biodiversity in an ecosystem (see Figure 10.1).

In the packet she sent home for Week 1 of the design activity, Ms. Cartwright challenged her students to think about how they could measure the biodiversity of birds in their backyard. She knew that scaffolding would be necessary for some of her students who might become frustrated if the activity was difficult for them at the beginning. She included

Figure 10.1. Standards Addressed in Bird-Feeding Program Design Activity

NGSS Science Standard

- MS-LS2-5: Evaluate competing design solutions for maintaining biodiversity and ecosystem services.

CCSSM Mathematics Practice Standards

- MP1: Make sense of problems and persevere in solving them.
- MP3: Construct viable arguments and critique the reasoning of others.
- MP4: Model with mathematics.

CCSS-ELA Standards

- W.6.2: Write informative/explanatory texts to examine a topic and convey ideas, concepts, and information through the selection, organization, and analysis of relevant content.

an identification guide for birds in her area and a log sheet. If she had been able to work with the students in the classroom, she might have allowed them to struggle at this stage; however, she felt it was important for the students to engage with the activity at the beginning, and she knew that these observation data would be necessary as they developed their designs.

Ms. Cartwright also posted a low-bandwidth video to social media her students used explaining how to perform observations at the same time each day, showing the local birds she observed and how to fill out the observation log. She knew that she could not use observation as her students worked as formative assessment, so she included an exit ticket where each student answered the following questions: "Describe one thing you learned from the activity this week," "What questions did you have as you worked on the activity?," and "What, if anything, was confusing about the activity?"

Ms. Cartwright knew that it was going to be challenging to keep her students engaged with the project, but she hoped that the success her students had with performing observations would motivate them as they later developed their bird-feeding program designs. She reviewed the exit tickets and considered how that information could guide revisions to her plans for the following week, which included a written response summarizing all of her students' comments, emphasizing what she liked about their learning so far, and addressing their questions.

In the packet she sent home for Week 2, Ms. Cartwright included a copy of her own completed observation sheet, small packets of three different types of bird food, and several potential bird feeder designs. She asked her students to choose at least one type of food, build at least one bird feeder from found or discarded materials around their home, and continue their observations for the next week. She also asked them to create a way of visually displaying their observation sheet data and to compare their observations to hers using that visual display. She again provided scaffolding with a low-bandwidth video on social media to show how she had set up her bird feeder this week, discussing design options she had considered, and showing several different types of graphs and figures to display her observation data.

Ms. Cartwright wanted to have her students interact in some way to help make up for not having a classroom group activity. She let her students know that she would make a booklet of their results from the first 2 weeks to send to the whole class and that she would place them in groups to develop helpful comments on other group members' designs. She again included the same exit ticket as formative assessment to guide her reflection as she planned future instruction.

In the packet sent home for Week 3, Ms. Cartwright again included her response to the exit tickets. Exit tickets were invaluable in the process and helped determine scaffolding and readjustment of the activity. She also provided booklets with all student designs and observation data displays. She asked students to comment on their group members' designs and data displays as well as on her own. She had her students analyze which birds fed at which feeder from the data provided by other students in their group. She also asked her students to redesign their own feeders based on their review of data from other groups and develop a feeding plan to maximize biodiversity in their location. She asked students to write reflections including what revisions they made to their design and how their peers' observation data informed those revisions.

As a final summative assessment, students prepared a presentation on their group's findings, presented their final design and results from their location, and discussed how feeding technology influenced local biodiversity. Students also completed a reflection on the design process. Ms. Cartwright created a booklet of the final presentations for the students and sent those home the following week.

CONNECTING EXAMPLES TO CORE COMPONENTS, AFFORDANCES, AND SUPPORTS

As we conclude this chapter, we hope you recognize how the examples provided align with the approach we share in this book. These examples addressed content standards from mathematics, science, and literacy (see Figure 10.1); utilized the design process with mathematical modeling through data collection and data display; and focused on students understanding systems, including how meal planning works within a family's preferences and economic situation and how bird feeding influences diversity in a small-scale ecosystem. Formative assessment (e.g., exit tickets, student designs, and reflections) was used for reflective teaching, group-worthy tasks were created, and complex instruction techniques were used to promote equity among learners.

These examples could have been taken further to address equity beyond the classroom. This would enable students to use what they are learning to critique their world and consider solutions to societal challenges. For example, the activity around designing a meal schedule could naturally lead into exploration of what foods are locally available and their relative cost. This could lead into discussions of how socioeconomics influences such availability and the impacts of healthy foods on different populations. In some of the contexts where our teacher colleagues work, fresh produce and healthy foods are hard to find and fast food restaurants predominate. These tend to be low-income urban and rural settings, and the impacts of food choice on population health could be explored as a foundation for designing solutions to these large-scale challenges.

Part III

MAKING IT YOUR OWN

CHAPTER 11

Linking an Integrative Series of Design and Mathematical Modeling Activities

Across the chapters in Part II, we shared many examples of our teacher colleagues delivering instruction guided by the approach we describe in this book. We discussed nuances related to how constraints are developed, how to scaffold student learning, and considerations when designing instruction across the artifact-to-process continuum, different content areas and grade levels, and instructional modalities. We hope this helps you consider how to use design thinking to integrate this approach into your own instruction. In Part III, we discuss how to make this approach your own by linking an integrative series of design-based learning activities and how to develop design-based learning activities starting from content standards, textbook problems, or your own past not-yet-design-based activities.

The examples we have provided so far have been relatively short series of activities or lessons across at most a few weeks of class time. In this chapter, we add to that by focusing on how longer thematically integrated series of lessons can be even more valuable for supporting deep learning across many content standards. Before providing examples of such integration, recall how problem-based, design-based, and project-based learning are related to the various types of integration described in Chapter 4.

PROBLEM-BASED, DESIGN-BASED, AND PROJECT-BASED INTEGRATION

The core idea behind problem-based learning is that students learn more authentically, meaning they remember and can apply what they learn to new situations better when they learn in the service of solving problems that are relevant to their lives. As described in the Introduction, design-based learning, the central focus of this book, is a special case or subset of problem-based learning where design thinking is utilized to build, test, and redesign potential solutions to real-world problems. As described in

Chapter 4, these approaches leverage the evidence base for content-specific integration, which is utilizing content standards from different domains, as those standards are linked in design-based activities. Approaching teaching and learning in this way enriches student understanding through methodological and process integration. This is true because real-world tasks and activities lead students to build competence using methods and processes from mathematical, scientific, and literacy practice domains. In other words, *doing* mathematics, science, and literacy in the pursuit of solving real-world problems enhances knowing in deeper, more integrative, and more broadly applicable ways. Learners come to see how practices and knowledge useful for solving one problem can be applied to solving other problems.

The activities and lessons we have described up to this point have been directed toward sets of mathematics, science, and literacy content standards. One of the important aspects of these activities is that they have been situated in real-world contexts and involved building potential solutions to problems encountered in those contexts. However, those contexts have been fairly specific, and the problems that we focused on occurred over relatively short instructional time frames. They have leveraged content and methodological integration in problem-based and design-based learning. While this can be tremendously valuable and is a great starting point, leveraging project-based learning and thematic integration can extend that value across months, semesters, and even years.

What do we mean by "fairly specific contexts," and what is the alternative? As an example of a specific context, remember Ms. Kowalski and Ms. Lundgren's design-based activities in Chapter 7 around how to keep a drink cold as a context for helping students better understand the nature of thermal energy and how it is transferred. This specific context could be expanded and many other specific contexts related to it under a more general context of energy efficiency or even a broad theme of human consumption and related social and environmental impacts. Specific contexts, such as efficient refrigeration or heating and cooling of rooms or buildings, could be directly connected to thermal transfer. The economic, social, and environmental impacts of energy generation could interconnect those specific contexts and explorations in a broader thematic unit that extended across months or semesters, covering large groupings of relevant content standards. That broader thematic unit could also leverage design thinking related to improving energy production, transport, and utilization. This could easily lead to a focus on societally relevant economic and environmental impacts that affect cultural and social groups differently, illuminating important equity considerations.

Developing months- or semester-long thematically integrated design-based learning units is not where we suggest you *begin* transforming your teaching practices with the approach we advocate in this book. However, once you have begun applying this approach and developed (and refined) several more specific design-based learning activities, it is a natural extension of our approach that allows students to learn many of the content standards you need them to master in authentically integrated contexts across the school year.

We have worked with teacher colleagues developing thematic units around farm and garden design, around rocket design, and around energy-efficient design. Topics associated with specific lessons within each of these units are shown in Figure 11.1. All of these units extended over multiple weeks, if not months, incorporated authentic literacy practices around reflection and persuasive data-based argumentation in writing and presentation, and included a focus on equity in the classroom and beyond. We know that these do not exhaust the possibilities of large thematic units through which our approach can be put into practice. We provide portions of one such unit in the remainder of this chapter as an example to help you think about how you might develop such a unit in your own teaching context.

ENERGY AND THE ENVIRONMENT THEMATIC UNIT

Energy and the environment can be a powerful theme for motivating many students and connecting deep learning across a wide range of design-based activities to address large numbers of content standards efficiently while also engaging students to address equity issues in their communities and beyond. Recent research has shown that climate change impacts lower-income and marginalized populations much more than others, and if

Figure 11.1. Thematic Units and Associated Topics.

Thematic Unit	Associated Lesson Topics
Farm and Garden Design	Ecological niches, biological diversity, and monocultures; pollinators; predator-prey relationships; genetic transmission and heredity; proportional and linear relationships; and human impacts
Rocket Design	Historical figures, civil rights, and social justice; Newtonian mechanics; graphing and modeling relationships; and computer coding
Energy-Efficient Design	Modeling with mathematics; energy transfer; energy utilization and efficiency in homes and schools; statewide energy production and usage; and environmental impacts of energy production

students in middle- and high-income countries receive climate education, it could lead to large reductions in atmospheric carbon dioxide (Cordero et al., 2020). Presenting the full scope of a months-long thematic unit such as this is beyond the scope of a single chapter, so what follows are some key considerations, moments, and lessons to demonstrate connections and help you think about how to create such units for your own classroom.

Setting the Stage

There are two broad categories of things to consider in terms of setting the stage for a thematic unit such as this: classroom culture and introducing the unit. The first needs to occur well before the unit starts and is about building a classroom culture where design-based activities, group-worthy tasks, and equitable learning are familiar to learners.

Beginning with smaller-scale design-based activities to acclimate yourself and your students to putting this approach into practice is important. Putting complex instruction practices into place in your classroom with group-worthy tasks and discussing the importance of everyone's contribution helps. Discussions of productive struggle and failure as a constructive and necessary feature of learning and assigning competence to learners who less frequently participate are critically important, as is your own application of the design process for reflective teaching practices. Once you have practiced this approach and have refined several smaller-scale design-based activities, then you are ready to consider a larger thematic unit.

In introducing a large thematic unit that will include many design-based activities, there are several considerations to keep in mind. You want to find ways to spark students' imaginations, connect to real-world issues they will find relevant to their lives, and set the stage for students to engage in active design thinking around those issues. We strongly encourage using multimedia for this, such as videos from the Internet, segments from movies, or pictures and videos you create yourself. If you want to support your students in considering equity beyond the classroom, as we hope you will, then consider including resources that prompt their thinking along those lines. For example, a video showing extreme weather events throughout the world could lead to a discussion of who such disasters most often directly impact.

The initial "hook" setting the stage for a thematic unit such as this should give students a view of the high-level issues at stake, such as climate change and the equity of extreme weather events, but also lead directly into consideration of foundational understandings that need to be developed to tackle those larger-scale problems. In this case, an understanding of the impact of climate change can lead into consideration of its causes, and from there into thinking about the importance of energy efficiency. We have found that starting with the contexts most closely related to students' lives and moving from there to broader contexts that impact them is a fruitful way to maintain relevance while building from foundational ideas to larger-scale societal issues. For instance, some of our teacher colleagues incorporated a recent fairly large-scale flooding event in their community as an entry into this kind of lesson.

Energy Efficiency at Home and on the Road

Energy efficiency–focused design activities that are close to home, pun intended, are a good place to start once the thematic unit has been introduced. You could even lead a student brainstorm around where energy is used in their everyday lives and let the group decide where the focus for a next activity might be. Some areas that might come up include heating/cooling bills, electricity costs to run video games and TVs, gas prices or fuel efficiency for vehicles, and likely many others as you see your students' creativity in action.

Ms. Kowalski and Ms. Lundgren's design-based activities in Chapter 7 around how to keep a drink cold are good initial activities to explore the nature of thermal energy and how it is transferred. Having already described that activity, we focus here on examples focused on understanding what impacts fuel efficiency in vehicles and how insulation impacts heating and cooling of homes. In addition to integrating mathematics, science, and literacy content standards, these lessons can provide a rich context for discussing equity beyond the classroom. How does driving older cars or living in older or less well-maintained housing impact energy costs for low-income households? Who is best able to upgrade insulation in their home? Are these possibly hidden costs of poverty that contribute to less upward social mobility?

Ms. Jones provided an example focused on energy efficiency of vehicles in her 7th-grade mathematics class. She wanted her students to understand how to use linear regression to solve real-world problems. She decided to have students design the procedure for a series of experiments to accurately measure rolling resistance for inflatable tires and show that tire inflation impacts rolling resistance and thereby fuel efficiency. *Rolling resistance* is a measure of the energy needed to overcome the friction between rolling tires and the road, and the Department of Energy estimates that 3% of fuel efficiency is wasted because of improperly inflated tires.

Ms. Jones introduced students to rolling resistance and the linear relationship between the force required to maintain a constant speed of a vehicle and the weight of the vehicle. She provided groups of students with wagons that had inflatable tires, a tire pressure gauge, a tire pump, various weights, a force meter, and a scale. She arranged to use the gym so that students could pull the wagons freely.

Ms. Jones did not provide directions for the experiments, but rather encouraged the students to try things, plot the data, and "see if their results make sense." Each group was required to collect data, explain how they ensured that they were getting accurate data, show their design of the experimental process that could be used by other students, plot their data, and analyze their results. Some important findings from students included the need to weigh the wagon and not just the weights they added, the need to develop a technique for pulling the wagon to get consistent force measurements, and the discovery that changing tire inflation changed the slope of their plots of force versus weight.

Energy efficiency can also be explored in students' homes. Students can design a plan working with their caregivers to reduce energy consumption at home. Reviewing energy bills can be a powerful component of design activities, as can the use of energy measuring devices such as the Kill-A-Watt. The Kill-A-Watt is a device that plugs into an electric outlet that you then plug a device into so that it measures both the power at any given time being drawn by the device and the total energy used over a period of time. We have found that energy and power can be concepts that students sometimes conflate, and that working with meters in an instructional unit on energy in the home can be a powerful way to focus on understanding units of energy consumed (kilowatt-hours and kilojoules), units of power (kilowatts), and how they relate to each other.

Energy Efficiency at School

Having built up foundational knowledge around thermal transfer and energy efficiency at home and on the road, a good place to take the exploration next is your own classroom and school. Here, we describe a design-based lesson focused on saving your school money so that there are more resources available for activities and events students care about.

Ms. Heller told her 8th-grade science students,

> The county school system has a $3 million shortfall this year, and will have to make budget cuts that will have a significant impact on students. These cuts would include discontinuing the use of computers except in a computer lab; discontinuing Physical Education, Music, and Art classes to save on teacher costs; requiring students to furnish ALL school supplies; and reducing sports programs to a total of two female sports and two male sports. To avoid these deep cuts, the student government is forming a grassroots movement to find ways for the school system to reduce energy costs. You are a member of this special student government committee and will present to decision-makers at the school and county level. You will need to provide them with knowledge to make educated decisions related to energy usage, costs, and sustainable

options. Your goal is to provide potential sustainability practices and solutions to your school district leadership, thus reducing energy costs and saving current programs. Your challenge will be to make suggestions related to the physical building and/or to the current usage practices. Your proposal will need to be based upon research, including but not limited to research conducted within your school and data gathered from the school district's board office. Explain the environmental and financial benefits for your decisions.

Ms. Heller provided students with several online resources (e.g., www .energy.gov) to get them started and encouraged them to search for additional resources. As groups worked toward potential design solutions, she identified promising ideas around which to create further explorations. One of these included monitoring the temperature inside and outside the single-pane windows in the classroom, calculating their energy efficiency, and comparing this with commercially available alternatives. Students were required to create a blueprint and project proposal, described as follows:

Your committee will need to decide on two projects that students can undertake to help make the school greener and reduce energy costs. Once you have decided on these projects, make a written plan, or a blueprint, explaining how the students can implement this plan as well as support needed from teachers, administrators, staff, the school board, and the community. Include charts, graphs, or drawings to help the audience understand your idea and why it makes sense. This plan will need to be approved by the school board, administrators, and teachers. You will need to create a proposal that includes research you have conducted to support the environmental and scientific reasons why this project will be beneficial and minimize the school's impact on the environment. Be able to defend why you chose this and the health and/or economic benefits of your idea.

Energy Efficiency at Scale

Once you have worked through energy efficiency on a small scale at home or on the road and extended that to the scale of a school building, a good place to go next is scaling up to consider renewable energy generation through the fabrication of wind turbines and statewide energy production and usage, which can be done using the excellent, and free, online book by David MacKay (2009). This allows you to directly make the connection to larger environmental impacts on a national or global scale. We have used a design competition activity (e.g., Regents of the University of Colorado, 2005) focused on building an efficient wind turbine to introduce ideas around renewable energy resources and as a foundation for a broader activity focused on designing a statewide energy plan.

PULLING IT ALL TOGETHER

In this chapter, we described how you can pull together your past experiences implementing our approach to efficiently address large groups of content standards in thematic design-based units. These units can build students prior knowledge and skills to address meaningful problems in their homes, transportation, schools, communities, and the world at large. They make ideal contexts for addressing larger societal issues of equity, and help build and maintain student engagement as they design plans to impact their own lives, family, and community.

CHAPTER 12

Address Any Content Standard and "Fix" Textbook Problems

Understanding our approach and adapting the resources described in this book for your students and your teaching context is a good start, but we have a more ambitious goal in mind. We hope you develop and refine your own thematically integrated series of design-based lessons and use design thinking to engineer your own lifelong learning. We believe these ideas can be powerful tools to transform your teaching and learning practices to help you support continuously deeper and more-equitable learning throughout the rest of your professional career. Similarly, we hope teacher educators and professional development providers will use these ideas to support other teachers doing the same. In support of those goals, we focus this final chapter on how to use design thinking to creatively prototype, test, and refine design-based activities (a) starting from scratch to address groups of content standards important for your learners and (b) starting from existing textbook problems or other not-yet-design-based learning activities. Throughout, we continue to emphasize reflective teaching and designing for equitable learning. We end this chapter describing how we hope you and other educators will use the design process to engineer lifelong learning and continuous improvement.

STARTING FROM STANDARDS

One of the consistent challenges the teachers we work with describe is deciding which standards to address through design-based learning activities; this challenge is even more pronounced for standards outside their primary content area (e.g., mathematics and literacy standards for science teachers). One approach we have found helpful is working through the design stages focused on the "problem" of creating deep and equitable learning. We touched on this idea in Leveraging the Design Process for Reflective Teaching (Chapter 3). Here we expand on that discussion, drawing on your now deeper understanding of our approach and focusing specifically on designing activities to promote deep and equitable learning.

Empathize, Define, Ideate, and Prototype

The process begins importantly with the Empathize stage, where you focus on what your students need. This focus should be informed by many different kinds of evidence. You need to consider your students' prior knowledge, of course, but also their socioemotional and metacognitive strengths and skills; their cultural, community, and home contexts; their attitudes and preferences; and their familiarity with classroom routines supportive of design-based learning and complex instruction. Your goal at this stage is to understand your learners so that the learning activities you design will motivate them to engage, connect with their prior knowledge, leverage their strengths, and fill gaps in their knowledge. To do that, you need to understand what they know, what their strengths are, what motivates them, what cultural and community contexts are relevant to them, and what they do not know yet.

With a solid understanding of your learners, you are ready to focus on the Define stage. Keep in mind the dynamic and iterative nature of the design process in every stage. You will cycle forward and backward many times as you continually design and redesign your instruction. What you know about your learners should inform which standards or foundational knowledge you focus on. Sometimes that foundational knowledge will be more about classroom culture and routines rather than, primarily, content.

If you have not previously done group-worthy tasks or design-based learning with your students, consider focusing initial efforts on one standard that your experience (or that of more experienced teacher colleagues) says is relatively easy for students to learn. This ideally happens toward the beginning of the year and might be review of one or more content standards from the previous year or other foundational knowledge you want to solidify for your learners. This allows you and your students to develop familiarity with whole-class and small-group routines, as well as familiarity with the processes and challenges of attending to status and equity, scaffolding productive struggle, and hands-on design-based learning activities where students build and create with more autonomy than in many classrooms. You want to get to a place where you have developed a classroom culture supporting struggle in learning that comes from the content more than from the process, and where the process and your instructional practices support that struggle being productive. Cycle back to the Empathize stage as needed to determine whether you are there yet, and include yourself as a learner because you are learning how to implement and fine-tune this instructional approach for your teaching and learning context.

Once you have established a classroom culture where you and your learners are comfortable with the approach generally, you are ready to tackle more-challenging content standards and objectives. Focus on content standards you are required to cover in your classroom and link those with meaningful content standards and practices from areas outside of your primary content area. This is the challenge we started this section focused on.

How do you identify which standards to focus design-based learning on? We suggest starting by looking for clusters of standards in your primary content area that you have previously taught together (see Figure 12.1). There are somewhat natural groupings of standards in each content area that fit well together. For example, the CCSSM link several standards together in clusters. The NGSS suggest connected mathematics and ELS/literacy standards for each science standard. These resources could be an excellent starting point for considering groups of standards for your design-based instruction.

Identifying a small cluster of standards, sometimes only two, in your primary content area is a good place to start. These should be standards

---

Figure 12.1. Example Mathematics and Science Content Standard Groupings

CCSS-M Mathematics Standards

- M.7.2: Recognize and represent proportional relationships between quantities.
- M.8.7: Graph proportional relationships, interpreting the unit rate as the slope of the graph, and compare two different proportional relationships represented in different ways (e.g., compare a distance–time graph to a distance–time equation to determine which of two moving objects has greater speed).
- M.8.9: Solve linear equations in one variable.

NGSS Science Standards

- S.7.ESS.7: Apply scientific principles to design a method for monitoring and minimizing a human impact on the environment.
- S.7.ESS.2: Develop a model to describe the cycling of water through earth's systems driven by energy from the sun and the force of gravity.
- MS-ETS1-1: Define the criteria and constraints of a design problem with sufficient precision to ensure a successful solution, taking into account relevant scientific principles and potential impacts on people and the natural environment that may limit possible solutions.
- MS-ETS1-2: Evaluate competing design solutions using a systematic process to determine how well they meet the criteria and constraints of the problem.
- MS-ETS1-3: Analyze data from tests to determine similarities and differences among several design solutions to identify the best characteristics of each that can be combined into a new solution to better meet the criteria for success.
- MS-ETS1-4: Develop a model to generate data for iterative testing and modification of a proposed object, tool, or process such that an optimal design can be achieved.

you are confident your students have foundational knowledge for but also present substantial opportunities for additional learning. In future revisions of your learning activities, you can focus on adding additional relevant standards that make sense to you. For example, science content standards that focus on design fit well together and can be integrated into mathematics activities. Addressing multiple standards with the same design-based learning activity is one way this approach adds power and efficiency to teaching and learning. Our teacher colleagues have been surprised when first incorporating this approach to find they have extra time at the end of the year to consolidate learning and prepare for standardized testing because they covered standards more efficiently than in past years.

Keep in mind the dynamic and iterative nature of the design process. We suggest you move forward at this point to Ideate and Prototype design-based activities to address content standards from your primary area before circling back to connect those activities to content standards from other areas. For example, if you are a science teacher and have identified specific science content standards to focus on, you can prototype a design-based activity focused on those standards and then come back to find relevant mathematics and literacy content standards and practices, perhaps using the suggestions in the NGSS. Often, incorporating those content areas will require you to come up with creative ways to change your initial prototype based on what you know of your learners. This likely means incorporating more-meaningful measurement, mathematical modeling, and literacy practices focused on communication and reflection.

Once you have prototyped a first lesson around content in your primary area, the next step is looking for relevant content standards and practices from other content areas. This can seem like a daunting task at first, especially if you have not previously had the time to review content standards documents outside your primary content area. You need to read those standards documents with a focus on what you know about your learners and the activities you have designed so far. It can be very helpful to share your thinking and activities with teacher colleagues from other content areas and solicit their ideas. If you have access to curriculum specialists, teacher educators, or professional development providers, they can provide useful input as well. The good news is that you will find considerable overlap, especially in practices, across content areas.

The design-based lessons we describe as examples throughout this book can help spark ideas for you as you work to integrate content standards and practices from beyond your primary content area. We suggest you cycle through Empathize, Define, Ideate, and Prototype several times while developing your design-based learning activity to cover multiple content standards and practices across science, mathematics, and literacy content areas before moving to the Test phase. Once you have what you feel is one or more solid prototypes of design-based learning activities, then you are ready for the Test phase, where you try it out in your classroom.

Test, Communicate, and Redesign

As you implement design-based learning in your classroom, remember what we have discussed related to Leveraging the Design Process for Reflective Teaching in Chapter 3 and Strategically Using Formative Assessment Data in Chapter 8. Build formative assessment into your teaching and learning practices so that you gain evidence of what seems to be working well and what needs tweaking or redesign. Work to become more conscious of your reflection-in-action as you are teaching and capture your thoughts and formative assessment evidence where you can focus on them more explicitly later with reflection-on-action. Here is where your teacher-as-learner role is critical, as you focus on what you are learning while you teach and leverage that to continuously improve classroom culture and student learning.

Our teacher colleagues consistently describe the value of sharing what is working, their challenges, and their thoughts about redesigning activities with their network of educators and even family and friends. One of our teacher colleagues said,

> We do it as teachers; that is the way that we learn. Now we have to help construct those connections for our students. The importance of collaboration for students, like we are collaborating now. We learn so much from each other, how much more so will students learn from each other in the classroom. (Ms. Harris)

Communicating what you are learning and the feedback you receive from others is powerful. Where you can, focus on growing your professional network in ways that allow you to do more of this. You can grow your professional network by reaching out to other teachers in your school to consider cross-disciplinary teaming opportunities; by reaching out to providers of professional development sessions you attend; by attending local, regional, or even national education conferences; and by seeking input from college and university faculty in education. We certainly would welcome hearing from you!

STARTING FROM EXISTING ACTIVITIES OR PROBLEMS

We have described how to start from scratch, so to speak, with standards to create and refine design-based learning activities. While this is a valuable approach, it is often easier to start from an existing activity or textbook problem that is not design-based and modify that activity or problem to fit the approach we advocate here. We mentioned in the previous section that when first looking for standards to drive design-based learning activities, a fruitful area to look to is groups of standards you have tended to teach together in the past. When you do that, you will naturally also find the activities and problems you used in the past to address those standards. These are likely not design based and do not incorporate all the aspects of the approach we describe in this book. However, those activities and problems can be rich foundations to build from.

You can use the design process to focus on redesigning those activities to fit this approach. You are essentially starting with a prototype and cycling back to redesign that prototype with the constraints and criteria driven by the details of our approach. We shared an example of this in Chapter 2, where Ms. Vaughan started with a project using cocktail sticks and candy to have students make 3D models and turned that into a rich design-based learning activity where students designed molecule-building kits to be used by other students.

You have almost certainly come across and used textbook problems to address relevant standards in your content area. Textbooks are sometimes structured specifically to address national content standards, so they are a rich foundation for initial ideas around which to build design-based learning activities. Dan Meyer (2013) provided a helpful example of doing this in his TED talk "Math Class Needs a Makeover." When you watch this talk, you will recognize a lot of overlap between our approach and what you see. Dan Meyer focuses on mathematics textbook problems and stripping away all measurements and substeps, using multimedia to show the problem in the real world, and engaging students in deciding what is worth measuring, how to measure it, and what to do with those measurements in order to design a solution. When we mentioned Dan Meyer in Chapter 3, we were focused on understanding what makes a group-worthy task, and here we focus on how to take textbook problems and turn them into rich design-based learning activities. While Dan Meyer focuses almost exclusively on mathematics, his ideas can be applied similarly to science textbooks.

Problems with many mathematics and science textbooks include that they provide too much structure and guidance, leading students through a set of specific steps from a well-defined problem to specific answers, while not providing enough (or any) real-world experience. Students do not develop autonomy and problem-solving self-efficacy when they follow clearly delineated steps to a predefined answer. Students do not connect their everyday experience to deep learning when they do not get real-world experience of the concepts they are learning. Learning can be made so much richer by removing those steps, situating the problem in a context made real through multimedia or actual physical objects in the classroom,

Figure 12.2. Sample Middle Grades Science Textbook Problem and Content Standard

What do these two changes have in common?

- mixing sand and water
- shaking up salad dressing

Select all that apply.

☐ Both are only physical changes.
☐ Both are chemical changes.
☐ Both are caused by heating.
☐ Both are caused by cooling.

NGSS Science Standard Addressed

- MS.PS1-2: Analyze and interpret data on the properties of substances before and after the substances interact to determine if a chemical reaction has occurred.

and allowing students to explore and develop the problem, solution procedures, and criteria for correctness. By doing that to a textbook problem, you can turn it into a design-based activity through which our approach can be implemented.

When reading science textbooks, it is common to see problems that are either step-by-step computation of formulas, as discussed by Dan Meyer (2013); word problems like the example shown in Figure 12.2; or reading comprehension problems like the one we will discuss later in this chapter. Careful analysis of the problem in Figure 12.2 reveals that in both cases they are physical changes involving mixing at least two different materials at room temperature. Physical changes are reversible, but the two physical changes mentioned in the book problem can be quite difficult to reverse. However, separating a mixture of sand and water is an example of purifying water and lends itself to engaging design activities. Students can identify potential purification techniques through Internet research, design their own versions of filters, and perform experiments to see how well their filters work. As students repeatedly mix sand into water and separate the mixture using different filtration techniques, they gain a deep understanding of physical mixtures and solutions.

Such a design-based approach allows for additional standards to be incorporated. Students could consider water shortages due to natural disasters and past and current geoscience projects, allowing incorporation of NGSS Science Standard MS.ESS.3.1: Construct a scientific explanation based on evidence for how the uneven distributions of Earth's mineral, energy, and groundwater resources are the result of past and current geoscience processes. For many students, access to clean water is impossible without additional filtration, and such activities are particularly authentic.

In addition to this science standard, the filtration project allows for mathematical analysis. How much water can the designed filters purify in an hour? How much sand can each filter remove before it needs replacement? What factors impact rate and longevity for various filters?

Another common textbook problem type requires students to read a selection of text and then answer questions to demonstrate comprehension. Rather than providing too much guidance, as in the last example, textbook problems of this sort do not provide direct experience related to what students are learning. Direct experience manipulating objects is one of the key affordances of design-based learning that helps students connect their own experience of the world to deep learning about systems.

As an example, Ms. Jones was teaching 6th-grade science, and in the past she had used a textbook problem like this to address the following standard: MS-PS-4: Construct and present arguments using evidence to support the claim that gravitational interactions are attractive and depend on the masses of interacting objects. This textbook problem asked students to read paragraphs defining gravity and inertia and relating those concepts to the motion of the earth and the moon. Students were then asked to write responses to a series of comprehension questions such as the following:

1. Explain why the Earth orbits the Sun. Include both of the terms gravity and inertia in your explanation.
2. Explain what would happen on Earth if our gravity were suddenly decreased (lowered)?
3. What would happen to the planets if the Sun disappeared?
4. Explain what would happen to the gravity on the Earth if its mass were suddenly decreased (lowered)?

Ms. Jones wanted to develop a design-based activity that addressed these same concepts, but before she could implement that activity, she learned that her school was going to be closed for at least the next week because of flooding in the area. Fortunately, the children at her school had Chromebooks and wireless hotspots so that they would be able to receive asynchronous instruction while they were home. She had planned to update her approach with design-based learning and mathematical modeling activities but wondered if she could do it using an online instruction modality.

Ms. Jones remembered playing lunar landing games when she was younger and searched the Internet to see if she could find an example. She found a number of options, including one at http://moonlander.seb .ly (see Figure 12.3). She realized that her students could play the game and design a *walkthrough*, or detailed set of instructions, to teach others how to master the landing process. She recognized that the lunar setting

Figure 12.3. Lunar Landing Online Video Game

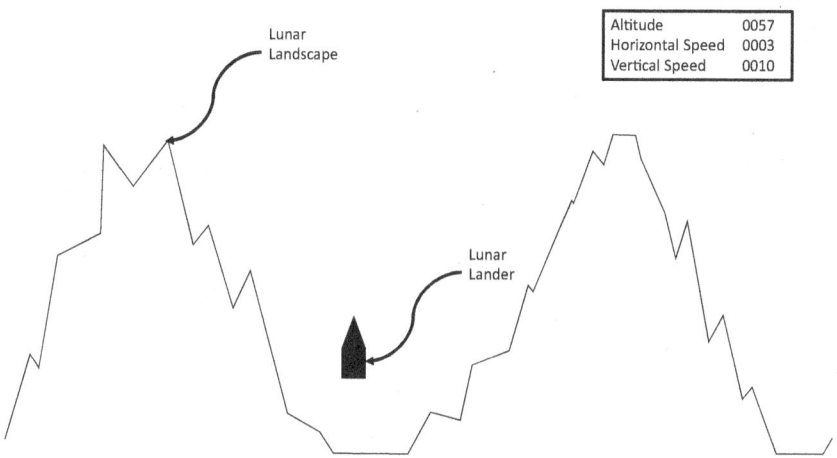

Source: Adapted from http://moonlander.seb.ly.

would facilitate natural discussions about the differences between gravity on the moon as compared to gravity on earth. Experiences playing the game would help her students connect more concretely to the concepts of gravity and inertia. She shared walkthroughs created by other students with the class to help them refine their own and asked them to write reflections comparing their walkthrough to another student's, explaining what they learned and how they applied that learning to improve their initial walkthrough. She ended the module by asking her students to design and present a written argument to support the following claim: The gravitational interaction between a lunar lander and the moon is attractive and depends on the mass of the moon.

Approaching textbook problems in this way allows for rich and authentic content integration. In addition to science and mathematics content, these are also authentic areas to integrate literacy standards and practices when you ask students to write and present reflections and scientific arguments describing what they are learning.

(RE)DESIGNING EFFECTIVE TEACHING PRACTICES

In Chapter 4, we described a professional development model we implemented with the following steps:

1. Identify mathematics and science knowledge gaps utilizing established standardized assessments.

2. Engage teachers in productive struggle as learners in integrative design-based tasks requiring that knowledge and including mathematical modeling for prediction prior to building and testing designs.
3. Require teachers to design, implement, and redesign design-based lessons addressing related knowledge gaps with their students.
4. Evaluate teacher learning through observations, interviews, and pre–post testing with standardized assessments; and redesign tasks to support further learning.

As we conclude this final chapter of the book, we want you to focus on using design thinking and a version of this model to guide your own lifelong learning and professional development.

You can use design thinking and reflective teaching practices to identify areas of opportunity for your own learning, whether those are related to content knowledge, instructional practices, specific approaches to instruction such as that described in this book, or anything else you feel relevant to your teaching practice. Then you can seek ways to pursue those opportunities through personal study, collaborative work with others in your professional network, formal professional development, action research and teacher inquiry, or anything else you feel can help you improve your practice.

It will be critical for you to put what you are learning into practice in your classroom so that you can fine-tune it for your teaching and learning context, refine it over time, and deepen your understanding through application in the real world. As you do so, you should utilize formative assessment to examine the impact of your instructional practices and reflect on your own learning to consider your next steps in continuously improving your instructional practices. Throughout, we hope you continue to focus on normalizing failure for yourself and your students, and on improving equity in your classroom so that all learners have access to deep and meaningful learning.

Conclusion
The Wicked Problem of Education for All

Thank you for including our book as part of your reflective teaching and lifelong learning. We truly believe that integrating our approach into your teaching and learning practice will be transformative for you and your students. As you know, this involves (a) finding problems and challenges that are meaningful to students, (b) engaging students in design thinking and mathematical modeling and scaffolding their progress while they build, test, and redesign solutions, and (c) attending to equity in and beyond the classroom while engaging in reflective teaching to continue to adjust scaffolding and keep students struggling productively to understand and improve their world.

As we conclude, we hope you have come to understand that design thinking is supremely adaptable. It is a process that applies authentically to design-based learning projects in and out of classrooms and to the design and redesign of teachers' instructional practices and professional development. It applies just as much to designing solutions to everyday problems as it does to "wicked problems" that the world faces today (Buchanan, 1992; Rittel & Webber, 1973; www.engineeringchallenges.org; www.wickedproblems.com). Wicked problems such as education, poverty, homelessness, climate change, and sustainability are in the headlines and impact all of us on a daily basis. As a teacher who cares about their students, you have in a very real sense dedicated much of your life to solving one of the most pivotal wicked problems: equitable education for all.

Wicked problems *seem* impossible to solve because we do not know all the factors involved. Those factors are interconnected and complex, and innovative multipronged solutions involving many different stakeholders are needed. Rittel and Webber (1973) described 10 characteristics of wicked problems:

1. There is no definitive formulation of a wicked problem.
2. Wicked problems have no stopping rule.
3. Solutions to wicked problems are not true-or-false, but good-or-bad.
4. There is no immediate and no ultimate test of a solution to a wicked problem.

5. Every solution to a wicked problem is a one-shot operation; because there is no opportunity to learn by trial-and-error, every attempt counts significantly.
6. Wicked problems do not have an enumerable (or exhaustively describable) set of potential solutions, nor is there a well-described set of permissible operations that may be incorporated into the plan.
7. Every wicked problem is essentially unique.
8. Every wicked problem can be considered to be a symptom of another problem.
9. The existence of a discrepancy representing a wicked problem can be explained in numerous ways. The choice of explanation determines the nature of the problem's resolution.
10. The planner has no right to be wrong.

Equity in education, whether focused inside a single classroom or more broadly across the education system, is a wicked and complex problem, one that you, as a teacher, engage in solving every day. Systems-level thinking is the only way to approach such problems, and design thinking is particularly effective for eliciting systems-level learning. Buchanan (1992) was the first to recognize the importance of design thinking in addressing wicked problems and emphasize using design thinking's iterative process to reframe problems in humancentric ways, create many ideas in brainstorming sessions, and adopt hands-on approaches in prototyping and testing. By now, you recognize how these ideas overlap with the design process, especially the Empathize-to-Define stages' focus on users' needs, the Ideate stage's focus on developing multiple solutions, and the Prototype-to-Test stages' focus on building and redesigning solutions.

We hope you will take the ideas you have learned throughout this book and integrate them into your work. Use this approach to: (a) engage students in design-based learning cycles to counter misconceptions and improve content knowledge, (b) engage yourself in reflective teaching practice cycles to build and refine design-based learning curricular units, (c) use design thinking to plan your own professional development to capitalize on areas of opportunity for continuous teaching practice improvement, and (d) use design thinking to develop improvements in everyday life, grand challenges, and wicked problems. As you do, please share your challenges and successes. Grow your professional network. Reach out to us. Seek additional ways to continue your lifelong learning in pursuit of improving the wicked problem of equitable education. We work to model and scaffold this mindset for and with our teacher colleagues. We have been gifted opportunities to see them do the same with us and with their students, and we hope that learners so enabled will contribute to the next generation of solutions the world so desperately needs.

References

Addleman, R. A., Brazo, C. J., Dixon, K., Cevallos, T., & Wortman, S. (2014). Teacher candidates' perceptions of debriefing circles to facilitate self-reflection during a cultural immersion experience. *The New Educator, 10*(2), 112–128.

Baker, K., Jessup, N. A., Jacobs, V. R., Empson, S. B., & Case, J. (2020). Productive struggle in action. *Mathematics Teacher: Learning & Teaching PK–12, 113*(5), 361–374.

Ball, D. L., & Forzani, F. M. (2011). Building a common core for learning to teach and connecting professional learning to practice. *American Educator, 35*(2), 17–21, 38–39.

Barlow, A. T., Gerstenschlager, N. E., Strayer, J. F., Lischka, A. E., Stephens, D. C., Hartland, K. S., & Willingham, J. C. (2018). Scaffolding for access to productive struggle. *Mathematics Teaching in the Middle School, 23*(4), 202–207.

Barrows, H. S. (1985). *How to design a problem-based curriculum for the preclinical years.* Springer.

Berry, A., & Milroy, P. (2002). Changes that matter. In J. J. Loughran, I. J. Mitchell, & J. Mitchell (Eds.), *Learning from teacher research* (pp. 196–221). Teachers College Press.

Bianchini, J. A. (1997). Where knowledge construction, equity, and context intersect: Student learning of science in small groups. *Journal of Research in Science Teaching, 34,* 1039–1065.

Bianchini, J. A. (1999). From here to equity: The influence of status on student access to and understanding of science. *Science Education, 83*(5), 577–601.

Blum, W. (2011). Can modeling be taught and learnt? Some answers from empirical research. In G. Kaiswer, W. Blum, R. Borromeo Ferri, & G. Stillman (Eds.), *Trends in teaching and learning of mathematical modeling* (pp. 15–30). Springer.

Boaler, J. (2006). How a detracked mathematics approach promoted respect, responsibility, and high achievement. *Theory Into Practice, 45*(1), 40–46.

Boaler, J. (2008). Promoting "relational equity" and high mathematics achievement through an innovative mixed-ability approach. *British Educational Research Journal, 34*(2), 167–194.

Bolyard, J., Curtis, R., Cairns, D., & Walker, A. (2018, April). *"Shoes of students . . . shoes of teachers": Experiencing and understanding productive struggle* [Paper presentation]. Annual meeting of the American Educational Research Association, New York, NY.

Bouillion, L. M., & Gomez, L. M. (2001). Connecting school and community with science learning: Real world problems and school–community partnerships as contextual scaffolds. *Journal of Research in Science Teaching, 38,* 878–898.

Buchanan, R. (1992). Wicked problems in design thinking. *Design Issues, 8*(2), 5–21.

Cairns, D., Curtis, R., Sierros, K., & Bolyard, J. (2018). Taking professional development from 2D to 3D: Design-based learning, 2D modeling, and 3D fabrication for authentic standards-aligned lesson plans. *Interdisciplinary Journal of Problem-Based Learning, 12*(2). https://docs.lib.purdue.edu/ijpbl/vol12/iss2/8/

Capra, F., & Luisi, P. L. (2014). *The systems view of life.* Cambridge University Press.

Center for Educational Policy. (2007). *Choice, changes, and challenges: Curriculum and instruction in the NCLB era.*

Chen, C., & Yang, Y. (2019). Revisiting the effects of project-based learning on students' academic achievement: A meta-analysis investigating moderators. *Educational Research Review, 26,* 71–81.

Cirillo, M., Pelesko, J. A., Felto-Kestler, M. D., & Rubel, L. (2016). Perspectives on modeling in school mathematics. In C. R. Hirsch & A. R. McDuffie (Eds.), *Annual perspectives in mathematics education: Mathematical modeling and modeling mathematics* (pp. 3–16). National Council of Teachers of Mathematics.

Cohen, E. G., & Lotan, R. A. (1997). Operation of status in the middle grades: Recent complications. In J. Szmatka, J. Skvoretz, & J. Berger (Eds.), *Status, network, and structure: Theory development in group processes* (pp. 222–240). Stanford University Press.

Cohen, E. G., & Lotan, R. A. (2014). *Designing groupwork: Strategies for the heterogeneous classroom* (3rd ed.). Teachers College Press.

Cohen, E. G., Lotan, R. A., Scarloss, B. A., & Arellano, A. R. (1999). Complex instruction in cooperative learning classrooms. *Theory Into Practice, 38*(2), 80–86.

Common Core State Standards Initiative. (2011). *Common Core State Standards for Mathematics.*

Common Core State Standards Writing Team. (2013). *Progressions for the Common Core State Standards in Mathematics (draft).* Institute for Mathematics and Education, University of Arizona.

Cordero, E. C., Centeno, D., & Todd, A. M. (2020). The role of climate change education on individual lifetime carbon emissions. *PLoS ONE, 15*(2), e0206266.

Curtis, R., Bolyard, J., Cairns, D., & Loomis, D. (2021, April). *"Step back and let them do that": Supporting middle-grades design-based teaching and learning* [Paper presentation]. Annual meeting of the American Educational Research Association, Virtual/Online.

Curtis, R., Bolyard, J., Cairns, D., Loomis, D. L., Mathew, S., & Watts, K. L. (2017a). Middle school math and science teachers engaged in STEM and literacy through engineering design. *Proceedings of the American Society for Engineering Education, USA, 2017,* 1–13.

Curtis, R., Bolyard, J., Cairns, D., Loomis, D. L., Mathew, S., & Watts, K. L. (2017b). Building middle school teacher mathematics and science content knowledge through engineering design. *Proceedings of the American Society for Engineering Education, USA, 2017,* 1–14.

Curtis, R., Bolyard, J., Cairns, D., Loomis, D. L., Mathew, S., & Watts, K. L. (2017c, April). *Teachers engaged in STEM and literacy (TESAL): Engineering design for middle school teaching and learning* [Paper presentation]. Annual meeting of the American Educational Research Association, San Antonio, TX.

Curtis, R., Bolyard, J., Cairns, D., Mathew, S., Loomis, D. L., & Watts, K. L. (2017, April). *Teachers as learners: A model to build teacher content knowledge through engineering design* [Poster presentation]. Annual meeting of the American Educational Research Association, San Antonio, TX.

Curtis, R., Cairns, D., & Bolyard, J. (2020). Understanding design, tolerating ambiguity, and developing middle school design-based lessons. *Proceedings of the American Society for Engineering Education, USA, 2020*, 1–25.

Curtis, R., Cairns, D., Bolyard, J., Loomis, D., Watts, K., Mathew, S., & Carte, M. (2016). Integrating STEM and literacy through engineering design: Evaluation of professional development for middle school math and science teachers. *Proceedings of the American Society for Engineering Education, USA, 2016*, 1–26.

Curtis, R., Cairns, D., Bolyard, J., & Walker, A. (2018, April). *Improving teacher mathematical content knowledge with integrative and sustained engineering design support* [Paper presentation]. Annual meeting of the American Educational Research Association, New York, NY.

Curtis, R., & Georgieva, Z. (2011, November). *Engineering research experiences and problem based learning in energy and environmental contexts* [Paper presentation]. Annual meeting of the West Virginia Science Teacher Association, Flatwoods, WV.

Curtis, R., Georgieva, Z., Cairns, D., & Solley, D. (2013, April). *Research experience for teachers (RET) in energy and the environment: Differential impact on science and mathematics teachers and implications for increased integration and teaming* [Paper presentation]. Annual meeting of the American Educational Research Association, San Francisco, CA.

Czerniak, C. (2007). Interdisciplinary science teaching. In S. Abell & N. Lederman (Eds.), *Handbook of research on science education* (pp. 537–559). Routledge.

Darling-Hammond, L., Flook, L., Cook-Harvey, C., Barron, B., & Osher, D. (2020). Implications for educational practice of the science of learning and development. *Applied Developmental Science, 24*(2), 97–140.

Dewey, J. (1929). *The quest for certainty*. Minton, Balch, & Co.

Dewey, J. (1933). *How we think: A restatement of the relation of reflective thinking to the educative process*. Houghton Mifflin.

Doerr, H. M., & English, L. D. (2006). Middle grade teachers' learning through students' engagement with modeling tasks. *Journal of Mathematics Teacher Education, 9*(1), 5–32.

Doerr, H. M., & Tinto, P. P. (2000). Paradigms for teacher-centered classroom-based research. In A. E. Kelly & R. A. Lesh (Eds.), *Handbook of research design in mathematics and science education* (pp. 403–429). Taylor & Francis Group.

Dym, C. L., Agogino, A. M., Eris, O., Frey, D. D., & Leifer, L. J. (2005). Engineering design thinking, teaching, and learning. *Journal of Engineering Education, 94*(1), 103–120.

Engle, R. A. (2006). Framing interactions to foster generative learning: A situation explanation of transfer in a community of learners classroom. *Journal of the Learning Sciences, 15*(4), 451–498.

Felton-Koestler, M. D. (2020, June). Teaching sociopolitical issues in mathematics teacher preparation: What do mathematics teacher educators need to know? *The Mathematics Enthusiast, 17*(2–3), 435–468.

Fortus, D., Krajcik, J., Dershimer, R., Marx, R. W., & Mamlok-Naaman, R. (2005). Design-based science and real-world problem-solving: Research report. *International Journal of Science Education, 27*(7), 855–879.

Forzani, F. M. (2014). Understanding "core practices" and "practice-based" teacher education: Learning from the past. *Journal of Teacher Education, 65,* 357–368.

Franke, M. L., Turrou, A. C., Webb, N. M., Ing, M., Wong, J., Shin, N., & Fernandez, C. (2015). Student engagement with others' mathematical ideas: The role of teacher invitation and support moves. *Elementary School Journal, 116*(1), 126–148.

Friday Institute for Educational Innovation. (2012). *Teacher efficacy and beliefs toward STEM survey.*

Furtak, E. M., Seidel, T., Iverson, H., & Briggs, D. C. (2012). Experimental and quasi-experimental studies of inquiry-based science teaching: A meta-analysis. *Review of Educational Research, 82*(3), 300–329.

Galbraith, P., & Stillman, G. (2006). A framework for identifying student blockages during transitions in the modelling process. *ZDM—The International Journal on Mathematics Education, 38*(2), 143–162.

Gann, C., Avineri, T., Graves, J., Hernandez, M., & Teague, D. (2016). Moving students from remembering to thinking: The power of mathematical modeling. In C. R. Hirsch & A. R. McDuffie (Eds.), *Annual perspectives in mathematics education: Mathematical modeling and modeling mathematics* (pp. 97–106). National Council of Teachers of Mathematics.

Geddis, A. N. (1996). Science teaching and reflection: Incorporating new subject-matter into teachers' classroom frames. *International Journal of Science Education, 18,* 249–265.

Georgieva, Z., Curtis, R., Saenz, T., Solley, D., & Cairns, D. (2013). Impact of attending a research experience for teachers program with international and societally relevant components. *Proceedings of the American Society for Engineering Education, USA, 2013,* 18403–18423.

Groshong, K. (2016). Different types of mathematical models. In C. R. Hirsch & A. R. McDuffie (Eds.), *Annual perspectives in mathematics education: Mathematical modeling and modeling mathematics* (pp. 17–24). National Council of Teachers of Mathematics.

Grossman, P., & McDonald, M. (2008). Back to the future: Directions for research in teaching and teacher education. *American Educational Research Journal, 45,* 184–205.

Gutiérrez, R. (2009). Framing equity: Helping students "play the game" and "change the game." *TODOS: Mathematics for ALL, 1*(1), 4–8.

Gutiérrez, R. (2012). Context matters: How should we conceptualize equity in mathematics education? In B. Herbel-Eisenmann, J. Choppin, D. Wagner, & D. Pimm (Eds.), *Equity in discourse for mathematics education: Theories, practices, and policies* (pp. 17–33). Springer.

Heaton, R. M., & Mickelson, W. T. (2002). The learning and teaching of statistical investigation in teaching and teacher education. *Journal of Mathematics Teacher Education, 5*, 35–59.

Hestenes, D., Wells, M., & Swackhamer, G. (1992). Force concept inventory. *The Physics Teacher, 30*(3), 141–158.

Hiebert, J., & Grouws, D. A. (2007). The effects of classroom mathematics teaching on students' learning. In F. K. Lester (Ed.), *Second handbook of research on mathematics teaching and learning* (pp. 371–404). Information Age Publishing.

Jaworski, B. (2006). Theory and practice in mathematics teaching development: Critical inquiry as a mode of learning in teaching. *Journal of Mathematics Teacher Education, 9*, 187–211.

Jett, C. C., & Cross, S. B. (2016). Teaching about diversity in black and white: Reflections and recommendations from two teacher educators. *The New Educator, 12*(2), 131–146.

Kaiser, G. (2017). The teaching and learning of mathematical modeling. In J. Cai (Ed.), *Compendium for research in mathematics education* (pp. 267–291). National Council of Teachers of Mathematics.

Kapur, M. (2008). Productive failure. *Cognition and Instruction, 26*(3), 379–424.

Kapur, M. (2010). Productive failure in mathematical problem solving. *Instructional Science, 38*(6), 523–550.

Kapur, M. (2011). A further study of productive failure in mathematical problem solving: Unpacking the design components. *Instructional Science, 39*(4), 561–579.

Kapur, M., & Bielaczyc, K. (2012). Designing for productive failure. *Journal of the Learning Sciences, 21*(1), 45–83.

Katz, S., & Stupel, M. (2016). Enhancing elementary-school mathematics teachers' efficacy beliefs: A qualitative action research. *International Journal of Mathematical Education in Science and Technology, 47*, 421–439.

Kolodner, J. L., Gray, J., & Fasse, B. B. (2003). Promoting transfer through case-based reasoning: Rituals and practices in Learning by Design™ classrooms. *Cognitive Science Quarterly, 3*, 119–170.

Kyei-Blankson, L. (2014). Training math and science teacher-researchers in a collaborative research environment: Implications for math and science education. *International Journal of Science and Mathematics Education, 12*, 1047–1065.

Lampert, M. (1990). When the problem is not the question and the solution is not the answer: Mathematical knowing and teaching. *American Educational Research Journal, 27*(1), 29–64.

Lampert, M. (2001). *Teaching problems and the problems in teaching.* Yale University Press.

Lazonder, A. W., & Harmsen, R. (2016). Meta-analysis of inquiry-based learning: Effects of guidance. *Review of Educational Research, 86*, 681–718.

Lesh, R., & Yoon, C. (2007). What is distinctive in (our views about) models and modelling perspectives on mathematics problem solving, learning, and teaching? In B. Werner, P. L. Galbraith, H. Henn, & M. Niss (Eds.), *Modelling and applications in mathematics education: The 14th ICMI Study* (pp. 153–160). Springer.

Lotan, R. A. (2003). Group-worthy tasks: Carefully constructed group learning activities can foster students' academic and social growth and help close the achievement gap. *Educational Leadership, 60*(6), 72–75.

Lotan, R. (2006). Teaching teachers to build equitable classrooms. *Theory Into Practice, 45*(1), 32–39.

MacKay, D. (2009). *Sustainable energy: Without the hot air.* UIT Cambridge Ltd. www.withouthotair.com

Manfra, M. M. (2019). Action research and systemic, intentional change in teaching practice. *Review of Research in Education, 43,* 163–196.

McDonald, M., Kazemi, E., & Kavanaugh, S. S. (2013). Core practices and pedagogies of teacher education: A call for a common language and collective activity. *Journal of Teacher Education, 64,* 378–386.

Meadows, D. H. (Eds.). (2008). *Thinking in systems: A primer.* Chelsea Green Publishing.

Mehalik, M. M., Doppelt, Y., & Schun, C. D. (2008). Middle-school science through design-based learning versus scripted inquiry: Better overall science concept learning and equity gap reduction. *Journal of Engineering Education, 97*(1), 71–85.

Meyer, D. (2009, September 8). *What I would do with this: Groceries.* https://blog.mrmeyer.com/2009/what-i-would-do-with-this-groceries/

Meyer, D. (2013, September 16). Math class needs a makeover. *TED-Ed talk.* https://www.ted.com/talks/dan_meyer_math_class_needs_a_makeover?

Miller, E. C., Reigh, E., Berland, L., & Krajcik, J. (2021). Supporting equity in virtual science instruction through project-based learning: Opportunities and challenges in the era of COVID-19. *Journal of Science Teacher Education, 32*(6), 642–663.

Nasir, N., Lee, C., Pea, R., & McKinney de Royston, M. (Eds.). (2020). *Handbook of the cultural foundations of learning.* Routledge.

National Academy of Engineering. (2008). 14 *Challenges for engineering in the 21st century.* http://www.engineeringchallenges.org/challenges.aspx

National Academies of Sciences, Engineering, and Medicine. (2018). *How people learn II: Learners, contexts, and cultures.* The National Academies Press. https://www.nap.edu/download/24783

National Council of Teachers of Mathematics. (2000). *Principles and standards for school mathematics.*

National Council of Teachers of Mathematics. (2014). *Principles to actions: Ensuring mathematical success for all.*

National Council of Teachers of Mathematics. (2016). *Annual perspectives in mathematics education: Mathematical modeling and modeling mathematics.*

National Governors Association & Council of Chief State School Officers. (2010). *Common Core Standards for Mathematics.* http://www.corestandards.org

National Research Council. (2006). *National science education standards.* National Academy.

National Research Council. (2012). *A framework for K–12 science education: Practices, cross-cutting concepts, and core ideas.* National Academy.

NGSS Lead States. (2013). *Next generation science standards: For states, by states.* The National Academies Press. http://www.nextgenscience.org

Ostorga, A. N., & Estrada, V. L. (2009). Impact of an action research instructional model: Student teachers as reflective thinkers. *Action in Teacher Education, 30*(4), 18–27.

Perrenet, J., & Terwel, J. (1997). Learning together in multicultural groups: A curriculum innovation. *Curriculum and Teaching, 12*(1), 31–44.

Piaget, J. (1960). The general problems of the psycho-biological development of the child. In *Discussions on child development* (Vol. 4, pp. 3–27). Tavistock.

Prince, M. (2004). Does active learning work? A review of the research. *Journal of Engineering Education, 93*(3), 223–232.

Pyzdrowski, L. J., Sun, L., Curtis, R., Miller, D., Winn, G., & Hensel, R. (2013). Readiness and attitudes as indicators for success in college calculus. *International Journal of Science and Mathematics Education, 11*(3), 529–554.

Raygoza, M. C. (2016). Striving toward transformational resistance: Youth participatory action research in the mathematics classroom. *Journal of Urban Mathematics Education, 9*(2), 122–152.

Regents of the University of Colorado. (2005). *Wind power! Designing a wind turbine.* https://www.teachengineering.org/activities/view/cub_energy2_lesson07_activity2

Rigney, J., Ferland, T., & Dana, N. F. (2019). Understanding teacher reflectivity in contemporary times: A (re)review of the literature. *The New Educator, 15*(4), 305–326.

Rittel, H. W., & Webber, M. M. (1973). Dilemmas in a general theory of planning. *Policy Sciences, 4*(2), 155–169.

Saderholm, J., Ronau, R., Brown, E. T., & Collins, G. (2010). Validation of the Diagnostic Teacher Assessment of Mathematics and Science (DTAMS) instrument. *School Science and Mathematics, 110*(4), 180–192.

Schön, D. (1983). *The reflective practitioner: How professionals think in action.* Basic Books.

Schön, D. (1987). *Educating the reflective practitioner: Toward a new design for teaching and learning in the professions.* Jossey-Bass.

Schwartz, D. L., & Martin, T. (2004). Inventing to prepare for future learning: The hidden efficiency of encouraging original student production in statistics instruction. *Cognition and Instruction, 22,* 129–184.

Skemp, R. R. (1971). *The psychology of learning mathematics.* Penguin.

Sokolowshi, A. (2015). The effects of mathematical modeling on students' achievement: Meta-analysis of research. *The International Academic Forum Journal of Education, 3*(1), 93–114.

Steffe, L. P. (1991). The constructivist teaching experiment: Illustrations and implications. In E. von Glasersfeld (Ed.), *Radical constructivism in mathematics education* (pp. 177–194). Kluwer Academic Publishers.

Stein, M. K., Smith, M. S., Henningsen, M. A., & Silver, E. A. (2009). *Implementing standards-based mathematics instruction: A casebook for professional development* (2nd ed.). National Council of Teachers of Mathematics.

Stigler, J. W., & Hiebert, J. (2004). *The teaching gap: Best ideas from the world's teachers for improving education in the classroom.* Free Press.

Stillman, G., & Brown, J. (2021). Modeling the phenomenon versus modeling the data set. *Mathematical Thinking and Learning, 23*(1), 1–26.

Thomas, W. J. (2000). *A review of project-based learning.* Autodesk Foundation.

Townsend, C., Slavit, D., & McDuffie, A. R. (2018). Supporting all learners in productive struggle. *Mathematics Teaching in the Middle School, 23*(4), 216–224.

Usak, M., Masalimova, A. R., Cherdymova, E. I., & Shaidullina, A. R. (2020). New playmaker in science education: COVID-19. *Journal of Baltic Science Education, 19*(2), 180–185.

Valentine, K. D., & Bolyard, J. (2018, April 13–17). *Creating a classroom culture that supports productive struggle: Pre-service teachers' reflections on teaching mathematics* [Paper presentation]. Annual Meeting of the American Educational Research Association, New York, NY.

von Glasersfeld, E. (1991). Abstraction, re-presentation, and reflection: An interpretation of experience and of Piaget's approach. In L. P. Steffe (Ed.), *Epistemological foundations of mathematical experience* (pp. 45–67). Springer.

Vygotsky, L. (1978). *Mind in society: The development of higher psychological processes*. Harvard University Press.

Warshauer, H. K. (2015a). Productive struggle in middle school mathematics classrooms. *Journal of Mathematics Teacher Education, 18*(4), 375–400.

Warshauer, H. K. (2015b). Strategies to support productive struggle. *Mathematics Teaching in the Middle School, 20*(7), 390–393.

Watanabe, M. (2012). *"Heterogenius" classrooms: Detracking math and science: A look at groupwork in action*. Teachers College Press.

Webb, N. M., Nemer, K. M., & Zuniga, S. (2002). Short circuits or superconductors? Effects of group composition on high-achieving students' science assessment performance. *American Educational Research Journal, 39*(4), 943–989.

Zawojewski, J. S. (2016). Mathematical modeling as a vehicle for STEM learning: Introduction. In C. R. Hirsch, & A. R. McDuffie (Eds.), *Annual perspectives in mathematics education: Mathematical modeling and modeling mathematics* (pp. 117–119). National Council of Teachers of Mathematics.

Index

About the Authors

Reagan Curtis (Ph.D. in Educational Psychology from the University of California, Santa Barbara) is Chester E. & Helen B. Derrick Endowed Professor in the School of Education and director of the Program Evaluation and Research Center at West Virginia University. Reagan works within a learning sciences perspective to focus on the evidence base for teaching and learning practices and within an evaluation research framework to develop and improve professional development programs.

Darran Cairns (Ph.D. in Materials Science and Engineering from the University of Birmingham, UK) is on the faculty in mechanical engineering at the University of Missouri at Kansas City, where he teaches graduate and undergraduate engineering and product design courses and is director of the School of Science and Engineering High School Math Academy. Darran brings deep engineering, science, and pedagogical expertise to supporting learners, as well as continuing a productive research agenda in materials science and engineering.

Johnna Bolyard (Ph.D. in Mathematics Education from George Mason University) is associate professor in the School of Education at West Virginia University, where she teaches undergraduate mathematics methods and content courses for teachers and graduate-level courses in mathematics education research and mathematics leadership. Johnna brings deep expertise in mathematics content and mathematics education pedagogy to support developing and practicing teachers. Her research focus is on the development of pre- and inservice mathematics teaching practice. In particular, she explores how teachers take up teaching practices that support the mathematical development of all students.